THE
SCIENTIFIC

MICHAEL FARADAY 1791–1867

EDWARD JENNER 1749–1823

THOMAS HUXLEY 1825–1895

GREGOR MENDEL 1822–1884

THE ART OF SCIENTIFIC INVESTIGATION

By

W. I. B. BEVERIDGE

Professor of Animal Pathology, University of Cambridge

" Scientific research is not itself a science;
it is still an art or craft."—W. H. GEORGE

Reprint of the 1957 Revised Edition, by W. W. Norton and Company, Inc.

The Art of Scientific Investigation

ISBN: 1-932846-05-0

Library of Congress Control Number: 2004111856

THE BLACKBURN PRESS
P. O. Box 287
Caldwell, New Jersey   07006
U.S.A.
973-228-7077
www.BlackburnPress.com

CONTENTS

*(The reference numbers throughout the book refer to
the numbers in the bibliography)*

LIST OF PLATES

PREFACE

Elaborate apparatus plays an important part in the science of to-day, but I sometimes wonder if we are not inclined to forget that the most important instrument in research must always be the mind of man. It is true that much time and effort is devoted to training and equipping the scientist's mind, but little attention is paid to the technicalities of making the best use of it. There is no satisfactory book which systematises the knowledge available on the practice and mental skills—the art—of scientific investigation. This lack has prompted me to write a book to serve as an introduction to research. My small contribution to the literature of a complex and difficult topic is meant in the first place for the student about to engage in research, but I hope that it may also interest a wider audience. Since my own experience of research has been acquired in the study of infectious diseases, I have written primarily for the student of that field. But nearly all the book is equally applicable to any other branch of experimental biology and much of it to any branch of science.

I have endeavoured to analyse the methods by which discoveries have been made and to synthesise some generalisations from the views of successful scientists, and also to include certain other information that may be of use and interest to the young scientist. In order to work this material into a concise, easily understandable treatise, I have adopted in some places a frankly didactic attitude and I may have over-simplified some of the issues. Nothing, however, could be further from my intentions than to be dogmatic. I have tried to deduce and state simply as many guiding principles of research as possible, so that the student may have some specific opinions laid before him. The reader is not urged to accept my views, but rather to look upon them as suggestions for his consideration.

Research is one of those highly complex and subtle activities that usually remain quite unformulated in the minds of those who practise them. This is probably why most scientists think that it is not possible to give any formal instruction in how to do research. Admittedly, training in research must be largely self-training, preferably with the guidance of an experienced scientist in the handling of the actual investigation. Nevertheless, I believe that some lessons and general principles can be learnt from the experience of others. As the old adage goes, " the wise man learns from the experience of others, the fool only from his own." Any training, of course, involves much more than merely being "told how". Practice is required for one to learn to put the precepts into effect and to develop a habit of using them, but it is some help to be told what are the skills one should acquire. Too often I have been able to do little more than indicate the difficulties likely to be met—difficulties which we all have to face and overcome as best we can when the occasion arises. Yet merely to be forewarned is often a help.

Scientific research, which is simply the search for new knowledge, appeals especially to people who are individualists and their methods vary from one person to another. A policy followed by one scientist may not be suitable for another, and different methods are required in different branches of science. However, there are some basic principles and mental techniques that are commonly used in most types of investigation, at least in the biological sphere. Claude Bernard, the great French physiologist, said :

> "Good methods can teach us to develop and use to better purpose the faculties with which nature has endowed us, while poor methods may prevent us from turning them to good account. Thus the genius of inventiveness, so precious in the sciences, may be diminished or even smothered by a poor method, while a good method may increase and develop it. . . . In biological sciences, the rôle of method is even more important than in the other sciences because of the complexity of the phenomena and countless sources of error." [15]

The rare genius with a flair for research will not benefit from instruction in the methods of research, but most would-be research workers are not geniuses, and some guidance as to how to go about research should help them to become productive earlier than they would if left to find these things out for themselves by the wasteful method of personal experience. A well-known scientist told me once that he purposely leaves his research students alone for some time to give them an opportunity to find their own feet. Such a policy may have its advantages in selecting those that are worthwhile, on a sink or swim principle, but to-day there are better methods of teaching swimming than the primitive one of throwing the child into water.

There is a widely held opinion that most people's powers of originality begin to decline at an early age. The most creative years may have already passed by the time the scientist, if he is left to find out for himself, understands how best to conduct research, assuming that he will do so eventually. Therefore, if in fact it is possible by instruction in research methods to reduce his non-productive probationary period, not only will that amount of time in training be saved, but he may become a more productive worker than he would ever have become by the slower method. This is only a conjecture but its potential importance makes it worth considering. Another consideration is the risk that the increasing amount of formal education regarded as necessary for the intending research worker may curtail his most creative years. Possibly any such adverse effect could be offset by instruction along the lines proposed.

It is probably inevitable that any book which attempts to deal with such a wide and complex subject will have many defects. I hope the shortcomings of this book may provoke others whose achievements and experience are greater than mine to write about this subject and so build up a greater body of organised knowledge than is available in the literature at present. Perhaps I have been rash in trying to deal with psychological aspects of research without having had any formal training in psychology; but I have been emboldened by the thought that a biologist venturing into psychology may be in no more danger of going seriously astray than would a psychologist or logician writing about biological research. Most books on the scientific method treat it from the logical or philosophical aspect. This one is more concerned with the psychology and practice of research.

I have had difficulty in arranging in a logical sequence the many diverse topics which are discussed. The order of the chapters on chance, hypothesis, imagination, intuition, reason and observation is quite arbitrary. The procedure of an investigation is epitomised in the second section of Chapter One. Trouble has been taken to collect anecdotes showing how discoveries have been made, because they may prove useful to those studying the ways in which knowledge has been advanced. Each anecdote is cited in that part of the book where it is most apt in illustrating a particular aspect of research, but often its interest is not limited to the exemplification of any single point. Other anecdotes are given in the Appendix. I apologise in advance for referring in several places to my own experience as a source of intimate information.

I sincerely thank many friends and colleagues to whom I am greatly indebted for helpful suggestions, criticism and references. The following kindly read through an early draft of the book and gave me the benefit of their impressions: Dr. M. Abercrombie, Dr. C. H. Andrewes, Sir Frederic Bartlett, Dr. G. K. Batchelor, Dr. A. C. Crombie, Dr. T. K. Ewer, Dr. G. S. Graham-Smith, Mr. G. C. Grindley, Mr. H. Lloyd Jones, Dr. G. Lapage, Sir Charles Martin, Dr. I. Macdonald, Dr. G. L. McClymont, Dr. Marjory Stephenson and Dr. D. H. Wilkinson. It must not be inferred, however, that these scientists endorse all the views expressed in the book.

PREFACE TO SECOND EDITION

It is most gratifying to be able to add now that the methods of research outlined in this book have received endorsement by a considerable number of scientists, both in reviews and in private communications. I have not yet met any serious disagreement with the main principles. Therefore it is now possible to offer the book with greater confidence.

I am deeply grateful to the many well-wishers who have written to me, some with interesting confirmation of views expressed in the book, and some drawing attention to minor errors. The alterations introduced in this second edition are for the most part minor revisions but the chapter on Reason has been partly rewritten.

Cambridge, July 1953.   W.I.B.B.

PREFACE TO THIRD EDITION

This edition differs only slightly from the previous one. The opportunity has been taken to make a few alterations, mostly of a minor nature, and add to the Appendix two good stories illustrating the rôle of chance.

Cambridge, September 1957.   W.I.B.B.

CHAPTER ONE

PREPARATION

" The lame in the path outstrip the swift
who wander from it."—FRANCIS BACON

Study

THE research worker remains a student all his life. Preparation
for his work is never finished for he has to keep abreast with
the growth of knowledge. This he does mainly by reading current
scientific periodicals. Like reading the newspapers, this study
becomes a habit and forms a regular part of the scientist's life.

The 1952 edition of the *World List of Scientific Periodicals*
indexes more than 50,000 periodicals. A simple calculation shows
this is equivalent to probably two million articles a year, or 40,000
a week, which reveals the utter impossibility of keeping abreast
of more than the small fraction of the literature which is most
pertinent to one's interest. Most research workers try to see
regularly and at least glance through the titles of the articles in
twenty to forty periodicals. As with the newspaper, they just skim
through most of the material and read fully only those articles
which may be of interest.

The beginner would be well advised to ask an experienced
research worker in his field which journals are the most important
for him to read. Abstracting journals are of limited value, if only
because they necessarily lag some considerable time behind the
original journals. They do, however, enable the scientist to cover
a wide range of literature and are most valuable to those who
have not access to a large number of journals. Students need
some guidance in ways of tracing references through indexing
journals and catalogues and in using libraries.

It is usual to study closely the literature dealing with the
particular problem on which one is going to work. However,
surprising as it may seem at first, some scientists consider that
this is unwise. They contend that reading what others have

written on the subject conditions the mind to see the problem in the same way and makes it more difficult to find a new and fruitful approach. There are even some grounds for discouraging an excessive amount of reading in the general field of science in which one is going to work. Charles Kettering, who was associated with the discovery of tetraethyl lead as an anti-knock agent in motor fuels and the development of diesel engines usable in trucks and buses, said that from studying conventional text-books we fall into a rut and to escape from this takes as much effort as to solve the problem. Many successful investigators were not trained in the branch of science in which they made their most brilliant discoveries: Pasteur, Metchnikoff and Galvani are well-known examples. A sheepman named J. H. W. Mules, who had no scientific training, discovered a means of preventing blowfly attack in sheep in Australia when many scientists had failed. Bessemer, the discoverer of the method of producing cheap steel, said:

"I had an immense advantage over many others dealing with the problem inasmuch as I had no fixed ideas derived from long established practice to control and bias my mind, and did not suffer from the general belief that whatever is, is right."

But in his case, as with many such "outsiders", ignorance and freedom from established patterns of thought in one field were joined with knowledge and training in other fields. In the same vein is the remark by Bernard that "it is that which we do know which is the great hindrance to our learning not that which we do not know." The same dilemma faces all creative workers. Byron wrote:

"To be perfectly original one should think much and read little, and this is impossible, for one must have read before one has learnt to think."

Shaw's quip "reading rots the mind" is, characteristically, not quite so ridiculous as it appears at first.

The explanation of this phenomenon seems to be as follows. When a mind loaded with a wealth of information contemplates a problem, the relevant information comes to the focal point of thinking, and if that information is sufficient for the particular problem, a solution may be obtained. But if that information is not sufficient—and this is usually so in research—then that mass of information makes it more difficult for the mind to conjure up original ideas, for reasons which will be discussed later. Further, some of that information may be actually false, in which case it presents an even more serious barrier to new and productive ideas.

Thus in subjects in which knowledge is still growing, or where the particular problem is a new one, or a new version of one already solved, all the advantage is with the expert, but where knowledge is no longer growing and the field has been worked out, a revolutionary new approach is required and this is more likely to come from the outsider. The scepticism with which the experts nearly always greet these revolutionary ideas confirms that the available knowledge has been a handicap.

The best way of meeting this dilemma is to read critically, striving to maintain independence of mind and avoid becoming conventionalised. Too much reading is a handicap mainly to people who have the wrong attitude of mind. Freshness of outlook and originality need not suffer greatly if reading is used as a stimulus to thinking and if the scientist is at the same time engaged in active research. In any case, most scientists consider that it is a more serious handicap to investigate a problem in ignorance of what is already known about it.

One of the most common mistakes of the young scientist starting research is that he believes all he reads and does not distinguish between the results of the experiments reported and the author's interpretation of them. Francis Bacon said :

" Read not to contradict and confute, nor to believe and take for granted . . . but to weigh and consider." [7]

The man with the right outlook for research develops a habit of correlating what is read with his knowledge and experience, looking for significant analogies and generalisations. This method of study is one way in which hypotheses are developed, for instance it is how the idea of survival of the fittest in evolution came to Darwin and to Wallace.

Successful scientists have often been people with wide interests.

3

Their originality may have derived from their diverse knowledge. As we shall see in a later chapter on Imagination, originality often consists in linking up ideas whose connection was not previously suspected. Furthermore, variety stimulates freshness of outlook whereas too constant study of a narrow field predisposes to dullness. Therefore reading ought not to be confined to the problem under investigation nor even to one's own field of science, nor, indeed, to science alone. However, outside one's immediate interests, in order to minimise time spent in reading, one can read for the most part superficially, relying on summaries and reviews to keep abreast of major developments. Unless the research worker cultivates wide interests his knowledge may get narrower and narrower and restricted to his own speciality. One of the advantages of teaching is that it obliges the scientist to keep abreast of developments in a wider field than he otherwise would.

It is more important to have a clear understanding of general principles, without, however, thinking of them as fixed laws, than to load the mind with a mass of detailed technical information which can readily be found in reference books or card indexes. For creative thinking it is more important to see the wood than the trees; the student is in danger of being able to see only the trees. The scientist with a mature mind, who has reflected a good deal on scientific matters, has not only had time to accumulate technical details but has acquired enough perspective to see the wood.

Nothing that has been said above ought to be interpreted as depreciating the importance of acquiring a thorough grounding in the fundamental sciences. The value to be derived from superficial and "skim" reading over a wide field depends to a large extent on the reader having a background of knowledge which enables him quickly to assess the new work reported and grasp any significant findings. There is much truth in the saying that in science the mind of the adult can build only as high as the foundations constructed in youth will support.

In reading that does not require close study it is a great help to develop the art of skim-reading. Skimming properly done enables one to cover a large amount of literature with economy of time, and to select those parts which are of special interest. Some styles of writing, of course, lend themselves more to skim-

4

ming than others, and one should not try to skim closely reasoned or condensed writing or any work which one intends to make the object of a careful study.

Most scientists find it useful to keep a card index with brief abstracts of articles of special interest for their work. Also the preparation of these abstracts helps to impress the salient features of an article in the memory. After reading quickly through the article to get a picture of the whole, one can go back to certain parts, whose full significance is then apparent, re-read these and make notes.

The recent graduate during his first year often studies some further subject in order better to fit himself for research. In the past it has been common for English-speaking research students to study German if they had no knowledge of that language and had already learnt French at school. In the biological sciences I think students would now benefit more from taking a course in biometrics, the importance of which is discussed in the next chapter. In the past it was important to be able to read German, but the output of Germany in the biological and medical sciences has been very small during the last ten years, and it does not seem likely to be considerable for some years to come. Scientists in certain other countries, such as Scandinavia and Japan, who previously often published in the German language, are now publishing almost entirely in English, which, with the vast expansion of science in America as well as throughout the British Commonwealth is becoming the international scientific language. Unless the student of biology has a special reason for wanting to learn German, I think he could employ his time more usefully on other matters until German science is properly revived. In this connection it may be worth noting the somewhat unusual view expressed by the great German chemist, Wilhelm Ostwald, who held that the research student should refrain from learning languages. He considered that the conventional teaching of Latin, in particular, destroys the scientific outlook.[67] Herbert Spencer has also pointed out that the learning of languages tends to increase respect for authority and so discourage development of the faculty of independent judgment, which is so important, especially for scientists. Several famous scientists—including Darwin and Einstein—had a strong distaste for Latin, probably

5

because their independent minds rebelled against developing the habit of accepting authority instead of seeking evidence.

The views expressed in the preceding paragraph on the possible harmful effect of learning languages are by no means widely accepted. However, there is another consideration to be taken into account when deciding whether or not to study a language, or for that matter any other subject. It is that time and effort spent in studying subjects not of great value are lost from the study of some other subject, for the active-minded scientist is constantly faced with what might be called the problem of competing interests : he rarely has enough time to do all that he would like to and should do, and so he has to decide what he can afford to neglect. Bacon aptly said that we must determine the relative value of knowledges. Cajal decries the popular idea that all knowledge is useful; on the contrary, he says, learning unrewarding subjects occupies valuable time if not actual space in the mind.[110] However, I do not wish to imply that subjects should be judged on a purely utilitarian basis. It is regrettable that we scientists can find so little time for general literature.

If the student cannot attend a course in biometrics, he can study one of the more easily understood books or articles on the subject. The most suitable that have come to my notice are those of G. W. Snedecor,[87] which deals with the application of statistics to animal and plant experimentation, and A. Bradford Hill,[16] which deals mainly with statistics in human medicine. Topley and Wilson's text-book of bacteriology contains a good chapter on the application of biometrics to bacteriology.[91] Professor R. A. Fisher's two books are classical works, but some people find them too difficult for a beginning.[39, 40] It is not necessary for the biologist to become an expert at biometrics if he has no liking for the subject, but he ought to know enough about it to avoid either undue neglect or undue respect for it and to know when he should consult a biometrician.

Another matter to which the young scientist might well give attention is the technique and art of writing scientific papers. The general standard of English in scientific papers is not high and few of us are above criticism in this matter. The criticism is not so much against the inelegance of the English as lack of clarity and accuracy. The importance of correct use of language lies not only in being able to report research well; it is with language that we do most of our thinking. There are several good short books and articles on the writing of scientific papers. Trelease[93] deals particularly with the technicalities of writing and editing and Kapp[55] and Allbutt[1] are mainly concerned with the writing of suitable English. Anderson[2] has written a useful paper on the preparation of illustrations and tables for scientific papers. I have found that useful experience can be gained by writing abstracts for publication. Thereby one becomes familiar with the worst faults that arise in reporting scientific work and at the same time one is subjected to a salutary discipline in writing concisely.

The scientist will find his life enriched and his understanding of science deepened by reading the lives and works of some of the great men of science. Inspiration derived from this source has given many young scientists a vision that they have carried throughout their lives. Two excellent recent biographies I can recommend are Dubos' *Louis Pasteur: Freelance of Science*[112] and Marquardt's *Paul Ehrlich.*[113] In recent years more and more attention is being given to the study of the history of science and every scientist ought to have at least some knowledge of this subject. It provides an excellent corrective to ever-increasing specialisation and broadens one's outlook and understanding of science. There are books which treat the subject not as a mere chronicle of events but with an insight which gives an appreciation of the growth of knowledge as an evolutionary process (e.g. [20, 25]). There is a vast literature dealing with the philosophy of science and the logic of scientific method. Whether one takes up this study depends upon one's personal inclinations, but, generally speaking, it will be of little help in doing research.

It is valuable experience for the young scientist to attend scientific conferences. He can there see how contributions to knowledge are made by building on the work of others, how papers are criticised and on what basis, and learn something of the personalities of scientists working in the same field as himself. It adds considerably to the interest of research to be personally acquainted with the authors of the papers one reads, or even merely to know what they look like. Conferences also

7

provide a good demonstration of the healthy democracy of science and the absence of any authoritarianism, for the most senior members are as liable to be criticised as is anyone else. Every opportunity should be taken to attend occasional special lectures given by eminent scientists as these can often be a rich source of inspiration. For instance, F. M. Burnet[19] said in 1944 that he had attended a lecture in 1920 by Professor Orme Masson, a man with a real feeling for science, who showed with superb clarity both the coming progress in atomic physics and the intrinsic delight to be found in a new understanding of things. Burnet said that although he had forgotten most of the substance of that lecture, he would never forget the stimulus it conveyed.

Setting about the Problem

In starting research obviously one has first to decide what problem to investigate. While this is a matter on which consultation with an experienced research worker is necessary, if the research student is mainly responsible for choosing his own problem he is more likely to make a success of it. It will be something in which he is interested, he will feel that it is all his own and he will give more thought to it because the responsibility of making a success of it rests on himself. It is wise for him to choose a subject within the field which is being cultivated by the senior scientists in his laboratory. He will then be able to benefit from their guidance and interest and his work will increase his understanding of what they are doing. Nevertheless, if a scientist is obliged to work on a given problem, as may be the case in applied research, very often an aspect of real interest can be found if he gives enough thought to it. It might even be said that most problems are what the worker makes them. The great American bacteriologist Theobald Smith said that he always took up the problem that lay before him, chiefly because of the easy access of material, without which research is crippled.[86] The student with any real talent for research usually has no difficulty in finding a suitable problem. If he has not in the course of his studies noticed gaps in knowledge, or inconsistencies, or has not developed some ideas of his own,

8

it does not augur well for his future as a research worker. It is best for the research student to start with a problem in which there is a good chance of his accomplishing something, and, of course, which is not beyond his technical capabilities. Success is a tremendous stimulus and aid to further progress whereas continued frustration may have the opposite effect.

After a problem has been selected the next procedure is to ascertain what investigations have already been done on it. Text-books, or better, a recent review article, are often useful as starting points, since they give a balanced summary of present knowledge, and also provide the main references. A text-book, however, is only a compilation of certain facts and hypotheses selected by the author as the most significant at the time of writing, and gaps and discrepancies may have been smoothed out in order to present a coherent picture. One must, therefore, always consult original articles. In each article there are references to other appropriate articles, and trails followed up in this way lay open the whole literature on the subject. Indexing journals are useful in providing a comprehensive coverage of references on any subject to within a year or so of the present, and where they cease a search is necessary in appropriate individual journals. *The Quarterly Cumulative Index Medicus, Zoological Record, Index Veterinarius* and the *Bibliography of Agriculture* are the standard indexing journals in their respective spheres. Trained librarians know how to survey literature systematically and scientists fortunate enough to be able to call on their services can obtain a complete list of references on any particular subject. It is advisable to make a thorough study of all the relevant literature early in the investigation, for much effort may be wasted if even only one significant article is missed. Also during the course of the investigation, as well as watching for new articles on the problem, it is very useful to read superficially over a wide field keeping constant watch for some new principle or technique that may be made use of.

In research on infectious diseases usually the next step is to collect as much firsthand information as possible about the actual problem as it occurs locally. For instance, if an animal disease is being investigated, a common procedure is to carry out field observations and make personal enquiries from farmers.

9

This is an important prerequisite to any experimental work, and occasionally investigators who have neglected it undertake laboratory work which has little relation to the real problem. Appropriate laboratory examination of specimens is usually carried out as an adjunct to this field work.

Farmers, and probably lay people generally, not infrequently colour their evidence to fit their notions. People whose minds are not disciplined by training often tend to notice and remember events that support their views and forget others. Tactful and searching enquiry is necessary to ascertain exactly what they have observed—to separate their observations from their interpretations. Such patient enquiry is often well repaid, for farmers have great opportunities of gathering information. The important discovery that ferrets are susceptible to canine distemper arose from an assertion of a gamekeeper. His statement was at first not taken seriously by the scientists, but fortunately they later decided to see if there was anything in it. It is said that for two thousand years the peasants of Italy have believed that mosquitoes were concerned with the spread of malaria although it was only about fifty years ago that this fact was established by scientific investigation.

It is helpful at this stage to marshal and correlate all the data, and to try to define the problem. For example, in investigating a disease one should try to define it by deciding what are its manifestations and so distinguish it from other conditions with which it may be confused. Hughlings Jackson is reported to have said : " The study of the causes of things must be preceded by the study of things caused." To show how necessary this is, there is the classical example of Noguchi isolating a spirochaete from cases of leptospiral jaundice and reporting it as the cause of yellow fever. This understandable mistake delayed yellow fever investigations (but the rumour that it led to Noguchi's suicide has no basis in fact). Less serious instances are not infrequently seen closer at hand.

The investigator is now in a position to break the problem down into several formulated questions and to start on the experimental attack. During the preparatory stage his mind will not have been passively taking in data but looking for gaps in the present knowledge, differences between the reports of different writers, inconsistencies between some observed aspect of the local problem and previous reports, analogies with related problems, and for clues during his field observations. The active-minded investigator usually finds plenty of scope for the formulation of hypotheses to explain some of the information obtained. From the hypotheses, certain consequences can usually be proved or disproved by experiment, or by the collection of further observational data. After thoroughly digesting the problem in his mind, the investigator decides on an experiment which is likely to give the most useful information and which is within the limitations of his own technical capacity and the resources at his disposal. Often it is advisable to start on several aspects of the problem at the same time. However, efforts should not be dispersed on too wide a front and as soon as one finds something significant it is best to concentrate on that aspect of the work.

As with most undertakings, the success of an experiment depends largely on the care taken with preliminary preparations. The most effective experimenters are usually those who give much thought to the problem beforehand and resolve it into crucial questions and then give much thought to designing experiments to answer the questions. A crucial experiment is one which gives a result consistent with one hypothesis and inconsistent with another. Hans Zinsser writing of the great French bacteriologist, Charles Nicolle, said:

" Nicolle was one of those men who achieve their successes by long preliminary thought before an experiment is formulated, rather than by the frantic and often ill-conceived experimental activities that keep lesser men in ant-like agitation. Indeed, I have often thought of ants in observing the quantity output of ' what-of-it ' literature from many laboratories. . . . Nicolle did relatively few and simple experiments. But every time he did one, it was the result of long hours of intellectual incubation during which all possible variants had been considered and were allowed for in the final tests. Then he went straight to the point, without wasted motion. That was the method of Pasteur, as it has been of all the really great men of our calling, whose simple, conclusive experiments are a joy to those able to appreciate them."[108]

Sir Joseph Barcroft, the great Cambridge physiologist, is said to have had the knack of reducing a problem to its simplest elements and then finding an answer by the most direct means. The general subject of planning research is discussed later under the title " Tactics ".

SUMMARY

One of the research worker's duties is to follow the scientific literature, but reading needs to be done with a critical, reflective attitude of mind if originality and freshness of outlook are not to be lost. Merely to accumulate information as a sort of capital investment is not sufficient.

Scientists tend to work best on problems of their own choice but it is advisable for the beginner to start on a problem which is not too difficult and on which he can get expert guidance.

The following is a common sequence in an investigation on a medical or biological problem. (a) The relevant literature is critically reviewed. (b) A thorough collection of field data or equivalent observational enquiry is conducted, and is supplemented if necessary by laboratory examination of specimens. (c) The information obtained is marshalled and correlated and the problem is defined and broken down into specific questions. (d) Intelligent guesses are made to answer the questions, as many hypotheses as possible being considered. (e) Experiments are devised to test first the likeliest hypotheses bearing on the most crucial questions.

CHAPTER TWO

EXPERIMENTATION

"The experiment serves two purposes, often independent one from the other: it allows the observation of new facts, hitherto either unsuspected, or not yet well defined; and it determines whether a working hypothesis fits the world of observable facts."—René J. Dubos.

Biological experiments

Science as we know it to-day may be said to date from the introduction of the experimental method during the Renaissance. Nevertheless, important as experimentation is in most branches of science, it is not appropriate to all types of research. It is not used, for instance, in descriptive biology, observational ecology or in most forms of clinical research in medicine. However, investigations of this latter type make use of many of the same principles. The main difference is that hypotheses are tested by the collection of information from phenomena which occur naturally instead of those that are made to take place under experimental conditions. In writing the last part of the previous chapter and the first part of this one I have had in mind the experimentalist, but there may be some points of interest in these also for the purely observational investigator.

An experiment usually consists in making an event occur under known conditions where as many extraneous influences as possible are eliminated and close observation is possible so that relationships between phenomena can be revealed.

The "controlled experiment" is one of the most important concepts in biological experimentation. In this there are two or more similar groups (identical except for the inherent variability of all biological material); one, the "control" group, is held as a standard for comparison, while the other, the "test" group, is subjected to some procedure whose effect one wishes to determine. The groups are usually formed by 'randomisation', that is to say, by assigning individuals to one group or the other by drawing lots or by some other means that does not involve human discrimination. The traditional method of experimentation is to have the groups as similar as possible in all respects except in the one variable factor under investigation, and to keep the experiment simple. " Vary one thing at a time and make a note of all you do." This principle is still widely followed, especially in animal experiments, but with the aid of modern statistical techniques it is now possible to plan experiments to test a number of variables at the same time.

As early as possible in an investigation, a simple crucial experiment should be carried out in order to determine whether or not the main hypothesis under consideration is true. The details can be worked out later. Thus it is usually advisable to test the whole before the parts. For example, before you try to reproduce a disease with a pure culture of bacteria it is usually wise to attempt transmission with diseased tissue. Before testing chemical fractions for toxicity, antigenicity or some other effect, first test a crude extract. Simple and obvious as this principle appears, it is not infrequently overlooked and consequently time is wasted. Another application of the same principle is that in making a first test of the effect of some quantitative factor it is usually advisable to determine at the outset whether any effect is produced under extreme conditions, for example, with a massive dose.

Another general principle of a rather similar kind is the process of systematic elimination. This method is well exemplified in the guessing game where a series of questions such as " animal, vegetable or mineral " is asked. One can often find the unknown more quickly by systematically narrowing down the possibilities than by making direct but blind guesses. This principle is used in weighing, when weights that are too heavy and too light are tried, and then the two extremes are gradually brought together. The method is especially useful in seeking an unknown substance by chemical means, but it also has many applications in various branches of biology. In investigating the cause of a disease, for instance, sometimes one eliminates the various alternatives until at last a narrow field is left for one to concentrate on.

In biology it is often good policy to start with a modest preliminary experiment. Apart from considerations of economy, it is seldom desirable to undertake at the outset an elaborate experiment designed to give a complete answer on all points. It is often better for the investigation to progress from one point to the next in stages, as the later experiments may require modification according to the results of the earlier ones. One type of preliminary experiment is the "pilot" experiment, which is often used when human beings or farm animals are the subjects. This is a small-scale experiment often carried out at the laboratory to get an indication as to whether a full-scale field experiment is warranted. Another type of preliminary experiment is the "sighting" experiment done to guide the planning of the main experiment. Take, for example, the case of an *in vivo* titration of an infective or toxic agent. In the sighting experiment dilutions are widely spaced (e.g. hundredfold) and few animals (e.g. two) are used for each dilution. When the results of this are available, dilutions less widely spaced (e.g. fivefold) are chosen just staggering the probable end-point, and larger groups of animals (e.g. five) are used. In this way one can attain an accurate result with the minimum number of animals.

The so-called "screening" test is also a type of preliminary experiment. This is a simple test carried out on a large number of substances with the idea of finding out which of them warrant further trial, for example, as therapeutic agents.

Occasionally quite a small experiment, or test, can be arranged so as to get a provisional indication as to whether there is anything in an idea which alone is based on evidence too slender to justify a large experiment. A sketchy experiment of this nature sometimes can be so planned that the results will be of some significance if they turn out one way though of no significance if the other way. However, there is a minimum below which it is useless to reduce the "set up" of even a preliminary experiment. If the experiment is worth doing at all it must be planned in such a way that it has at least a good chance of giving a useful result. The young scientist is often tempted through impatience, and perhaps lack of resources, to rush in and perform ill-planned experiments that have little chance of giving significant results. Sketchy experiments are only justifiable when preliminary to more elaborate experiments planned to give a reliable result. Each stage of the investigation must be established beyond reasonable doubt before passing on to the next, or else the work may be condemned, quite properly, as being "sloppy".

The essence of any satisfactory experiment is that it should be reproducible. In biological experiments it not infrequently happens that this criterion is difficult to satisfy. If the results of the experiment vary even though the known factors have not been altered, it often means that some unrecognised factor or factors is affecting the results. Such occurrences should be welcomed, because a search for the unknown factor may lead to an interesting discovery. As a colleague remarked to me recently: " It is when experiments go wrong that we find things out." However, first one should see if a mistake has been made, as a technical error is the most common explanation.

In the execution of the experiment it is well worth while taking the greatest care with the essential points of technique. By taking great pains and paying careful attention to the important details the originator of a new technical method sometimes is able to obtain results which other workers, who are less familiar with the subject or less painstaking, have difficulty in repeating. It is in this connection that Carlyle's remark that genius is an infinite capacity for taking pains is true. A good example is provided by Sir Almroth Wright's selection of the Rawlings strain of typhoid bacillus when he introduced vaccination against that disease. Only quite recently, since certain techniques have become available, has it been found that the Rawlings strain was an exceptionally good strain for use in making vaccine. Wright had carefully chosen the strain for reasons which most people would have considered of no consequence. Theobald Smith, one of the few really great bacteriologists, said of research :

"It is the care we bestow on apparently trifling, unattractive and very troublesome minutiae which determines the result." [17]

Some discrimination, however, should be used, for it is possible to waste time in elaborating unnecessary detail on unimportant aspects of the work.

The careful recording of all details in experimental work is an elementary but important rule. It happens surprisingly often that one needs to refer back to some detail whose significance one did not realise when the experiment was carried out. The notes kept by Louis Pasteur afford a beautiful example of the careful recording of every detail. Apart from providing an invaluable record of what is done and what observed, note-taking is a useful technique for prompting careful observation.

The experimenter needs to have a proper understanding of the technical methods he uses and to realise their limitations and the degree of accuracy attainable by each. It is essential to be thoroughly familiar with laboratory methods before using them in research and to be able to obtain consistent and reliable results. There are few methods that cannot at times go wrong and give misleading results and the experimenter should be able to detect trouble of this nature quickly. Where practicable, estimations and titrations of crucial importance should be checked by a second method. The scientist must also understand his apparatus. Modern complicated apparatus is often convenient but it is not always foolproof, and experienced scientists often tend to avoid it because they fear it may give misleading results.

Difficulties often arise in organising experiments with subjects over which there is only limited control—human beings or valuable farm animals. Unless the basic needs of the controlled experiment can be satisfied it is better to abandon the attempt. Such a statement may appear self-evident, but not infrequently investigators find the difficulties too great and compromise on some arrangement that is useless. Large numbers in no way offset the necessity of a satisfactory control group. The outstanding illustration is supplied by the story of B.C.G. vaccination in children. This procedure was introduced twenty-five years ago and was then claimed to protect people against tuberculosis; but although a large number of experiments have since been carried out, there is still to-day controversy as to its value in preventing the disease in people of European stock. Most of the experiments have proved nothing because the controls were not strictly comparable. The review on B.C.G. vaccination by Professor G. S. Wilson provides a good lesson in the difficulties and pitfalls of experimentation. He concludes:

"These results show how important it is when carrying out a controlled investigation on human subjects to do everything possible to ensure that the vaccinated and control children are similar in every respect, including such factors as age, race, sex, social, economic and housing conditions, intellectual level and co-operativeness of the parents, risk of exposure to infection, attendance at infant welfare or other clinics and treatment when ill."[106]

Professor Wilson has pointed out to me in conversation that unless decisive experiments are done before an alleged remedy is released for use in human medicine, it is almost impossible subsequently to organise an experiment with untreated controls, and so the alleged remedy becomes adopted as a general practice without anyone knowing if it is really of any use at all. For example, Pasteur's rabies treatment has never been proved by proper experiment to prevent rabies when given to persons after they are bitten and some authorities doubt if it is of any value, but it is impossible now to conduct a trial in which this treatment is withheld from a control group of bitten persons.

Sometimes it is a necessary part of a field experiment to keep the groups in different surroundings. In such experiments one cannot be sure that any differences observed are due to the particular factor under scrutiny and not to other variables associated with the different environments. This difficulty can sometimes be met by replicating both test and control groups so that any effects due to environment will be exposed and perhaps cancel out. If variables which are recognised but thought to be extraneous cannot be eliminated, it may be necessary to employ a series of control groups, or carry out a series of experiments, in order to isolate experimentally each known difference between the two populations being compared.

Whenever possible the results of experiments should be assessed by some objective measurement. However, occasionally this cannot be done, as for instance where the results concern the severity of clinical symptoms or the comparison of histological changes. When there is a possibility of subjective influences affecting the assessment of results, it is important to attain objectivity by making sure that the person judging the results does not know to which group each individual belongs. No matter how objectively minded the scientist may believe himself to be, it is very difficult to be sure that his judgment may not be subconsciously biased if he knows to which group the cases belong when he is judging them. The conscientious experimenter, being aware of the danger, may even err by biasing his judgment in the direction contrary to the expected result. Complete intellectual honesty is, of course, a first essential in experimental work.

When the experiment is complete and the results have been assessed, if necessary with the aid of biometrics, they are interpreted by relating them to all that is already known about the subject.

Planning and assessing experiments

Biometrics, or biostatistics, the application of the methods of mathematical statistics to biology, is a comparatively new branch of science and its importance in research has only lately won general recognition. Books dealing with this subject have been mentioned in Chapter One and I do not intend to do more here than call attention to a few generalities and stress the need for the research worker to be acquainted at least with the general principles. Some knowledge of statistical methods is necessary for any form of experimental or observational research where numbers are involved, but especially for the more complex experiments where there is more than one variable.

One of the first things which the beginner must grasp is that statistics need to be taken into account when the experiment is being planned, or else the results may not be worth treating statistically. Therefore biometrics is concerned not only with the interpretations of results but also with the planning of experiments. It is now usually taken as including, besides the purely statistical techniques, also the wider issues involved in their application to experimentation such as the general principles of the design of experiments and the logical issues concerned. Sir Ronald Fisher, who has done so much to develop biometrical methods, discusses these topics in his book, *The Design of Experiments*.[39]

In selecting control and test groups, logic and common sense have first to be satisfied. A common fallacy, for instance, is to compare groups separated by time—the data of one year being compared to data obtained in previous years. Evidence obtained in this way is never conclusive, though it may be usefully suggestive. "If when the tide is falling you take out water with a twopenny pail, you and the moon can do a great deal." In biological investigations there may be many unsuspected factors that influence populations separated by time or geographically. When general considerations have been satisfied, statistical methods are used to decide on the necessary size of the groups, to select animals according to weight, age, etc. and, while taking these particulars into account, to distribute the animals into groups without sacrificing the principle of random selection.

No two groups of animals or plants are ever exactly similar, owing to the inherent variability of biological material. Even though great pains are taken to ensure that all individuals in both groups are nearly the same in regard to sex, age, weight, breed, etc., there will always be variation that depends on factors not yet understood. It is essential to realise the impossibility of obtaining exactly similar groups. The difficulty must be met by estimating the variability and taking it into account when assessing the results. Within reasonable limits it is desirable to choose the animals for an experiment showing little variability one with another, but it is not essential to go to great lengths to achieve this. Its purpose is to increase the sensitivity of the experiment, but this can be done in other ways, such as by increasing the numbers in the groups. There are mathematical techniques for making corrections in certain cases for differences between individuals or groups.

Another method of meeting the difficulty of variability in experimental animals is by "pairing": the animals are arrayed in pairs closely resembling each other (perhaps pairs of twins or litter mates). Each animal is compared only with its fellow and thus a series of experimental results is obtained. By using identical twins one can often effect great economy in numbers, which is important in investigations on animals that are expensive to buy and keep. Experiments carried out in New Zealand on butterfat yield showed that as much information was obtained per pair of identical twin cows as from two groups each of 55 cows. In experiments with growth rates, identical twins were about 25 times more useful than ordinary calves.[4]

When testing out a procedure for the first time it is often impossible to estimate in advance how many animals are required to ensure a decisive result. If expensive animals are involved economy may be effected by doing a test first with a few animals and repeating the test until the accumulated results are sufficient to satisfy statistical requirements.

One of the basic conceptions in statistics is that the individuals in the group under scrutiny are a sample of an infinitely large, hypothetical population. Special techniques are available for random sampling and for estimating the necessary size of the sample for it to be representative of the whole. The number required in the sample depends on the variability of the material and on the degree of error that will be tolerated in the results, that is to say, on the order of accuracy required.

Fisher considers that in the past there has been too much emphasis placed on the importance of varying only one factor at a time in experimentation and shows that there are distinct advantages in planning experiments to test a number of variables at the same time. Appropriate mathematical techniques enable several variables to be included in the one experiment, and this not only saves time and effort, but also gives more information than if each variable were treated separately. More information is obtained because each factor is examined in the light of a variety of circumstances, and any interaction between the factors may be detected. The traditional method of experimental isolation of a single factor often involves a somewhat arbitrary definition of that factor and the testing of it under restricted, unduly simplified circumstances. Complex, multiple factor experiments, however, are not so often applicable to work with animals as to work with plants, although they can be used with advantage in feeding trials where various combinations of several components in the ration are to be tested.

Statistics, of course, like any other research technique, has its uses and its limitations and it is necessary to understand its proper place and function in research. It is mainly valuable in testing an hypothesis, not in initiating a discovery. Discoveries may originate from taking into consideration the merest hints, the slightest differences in the figures between different groups, suggesting something to be followed up; whereas statistics are usually concerned with carefully pre-arranged experiments set up to test an idea already born. Also, in trying to provide sufficient data for statistical analysis, the experimenter must not be tempted to do so at the expense of accurate observation and of care with the details of the experiment.

The use of statistics does not lessen the necessity for using common sense in interpreting results, a point which is sometimes forgotten. Fallacy is especially likely to arise in dealing with field data in which there may be a significant difference between two groups. This does not necessarily mean that the difference is caused by the factor which is under consideration because possibly there is some other variable whose influence or importance has not been recognised. This is no mere academic possibility, as is shown for example by the confusion that has arisen in many experiments with vaccination against tuberculosis, the common cold and bovine mastitis. Better hygienic measures and other circumstances which may influence the results are often coupled with vaccination. Statistics may show that people who smoke do not on the average live as long as people who do not smoke but that does not necessarily mean that smoking shortens life. It may be that people who do not smoke take more care of their health in other and more important ways. Such fallacies do not arise in well designed experiments where the initial process of randomisation ensures a valid comparison of the groups.

The statistician, especially if he is not also a biologist, may be inclined to accept data given him for analysis as more reliable than they really are, or as being estimated to a higher degree of accuracy than was attempted. The experimenter should state that measurements have been made only to the nearest centimetre, gram or whatever was the unit. It is helpful for the statistician to have had some personal experience of biological experimentation and he ought to be thoroughly familiar with all aspects of experiments on which he is advising. Close co-operation between the statistician and the biologist can often enable enlightened common sense to by-pass a lot of abstruse mathematics.

Occasionally scientific reports are marred by the authors giving their results only as averages. Averages often convey little information and may even be misleading. The frequency distribution should be given and some figures relating to individuals are often helpful in giving a complete picture. Graphs also can be misleading and the data on which they are based needs to be examined critically. If the plotted points on a graph are not close together—that is, if the observations have not been made at frequent intervals—it is not always justifiable to connect them with straight or curved lines. Such lines may not represent the true position, for one does not know what actually occurred in the interval. There may, for instance, have been an unsuspected rise and fall.

Misleading experiments

Some of the hazards associated with the use of reason, hypothesis and observation in research are discussed in the appropriate chapters of this book. As a corrective to any tendency to put excessive faith in experimentation, it is as well here to remind the reader that experiments also can at times be quite misleading. The most common cause of error is a mistake in technique. Reliance cannot be placed on results unless the experimenter is thoroughly competent and familiar with the technical procedures he uses. Even in the expert's hands technical methods have to be constantly checked against known " positive " and " negative " specimens. Apart from technical slips, there are more subtle reasons why experiments sometimes " go wrong ".

John Hunter deliberately infected himself with gonorrhoea to find out if it was a distinct disease from syphilis. Unfortunately the material he used to inoculate himself contained also the syphilis organism, with the result that he contracted both diseases and so established for a long time the false belief that both were manifestations of the same disease. Needham's experiments with flasks of broth led himself and others to believe that spontaneous generation was possible. Knowledge at the time was insufficient to show that the fallacy arose either from accidental contamination or insufficient heating for complete sterilisation. In recent years we have seen an apparently well-conducted experiment prove that patulin has therapeutic value against the common cold. Statistical requirements were well satisfied. But no one since has been able to show any benefit from patulin and why it seemed to be efficacious in the first experiment remains a mystery.[24]

When I saw a demonstration of what is known as the Mules operation for the prevention of blowfly attack in sheep, I realised its significance and my imagination was fired by the great potentialities of Mules' discovery. I put up an experiment involving thousands of sheep and, without waiting for the results, persuaded colleagues working on the blowfly problem to carry out experiments elsewhere. When about a year later, the results became available, the sheep in my trial showed no benefit from the operation. The other trials, and all subsequent ones, showed that the operation conferred a very valuable degree of protection and no satisfactory explanation could be found for the failure of my experiment. It was fortunate that I had enough confidence in my judgment to prevail upon my colleagues to put up trials in other parts of the country, for if I had been more cautious and awaited my results they would probably have retarded the adoption of the operation for many years.

Several large-scale experiments in the U.S.A. proved that immunisation greatly reduced the incidence of influenza in 1943 and again in 1945, yet in 1947 the same type of vaccine failed. Subsequently it was found that this failure was due to the 1947 strain of virus being different from those current in earlier years and used in making the vaccine.

It is not at all rare for scientists in different parts of the world to obtain contradictory results with similar biological material. Sometimes these can be traced to unsuspected factors, for instance, a great difference in the reactions of guinea-pigs to diphtheria toxin was traced to a difference in the diets of the animals. In other instances it has not been possible to discover the cause of the disagreement despite a thorough investigation. In Dr. Monroe Eaton's laboratory in the United States influenza virus can be made to spread from one mouse to another, but in Dr. C. H. Andrewes' laboratory in England this cannot be brought about, even though the same strains of mice and virus, the same cages and an exactly similar technique are used.

We must remember that, especially in biology, experimental results are, strictly speaking, only valid for the precise conditions under which the experiments were conducted. Some caution is necessary in drawing conclusions as to how widely applicable are results obtained under necessarily limited sets of circumstances.

Darwin once said half seriously, " Nature will tell you a direct lie if she can." Bancroft points out that all scientists know from experience how difficult it often is to make an experiment come out correctly even when it is known how it ought to go. Therefore, he says, too much trust should not be put in an experiment done with the object of getting information.[10]

The examples quoted are experiments which gave results that were actually "wrong" or misleading. Fortunately they are exceptional. Commoner, however, is the failure of an experiment to demonstrate something because the exact conditions necessary are not known, such as Faraday's early repeated failures to obtain an electric current by means of a magnet. Such experiments demonstrate the well-known difficulty of proving a negative proposition, and the folly of drawing definite conclusions from them is usually appreciated by scientists. It is said that some research institutes deliberately destroy records of "negative experiments", and it is a commendable custom usually not to publish investigations which merely fail to substantiate the hypothesis they were designed to test.

SUMMARY

The basis of most biological experimentation is the controlled experiment, in which groups, to which individuals are assigned at random, are comparable in all respects except the treatment under investigation, allowance being made for the inherent variability of biological material. Two useful principles are to test the whole before the part, and to eliminate various possibilities systematically. In the execution of an experiment close attention to detail, careful note-taking and objectivity in the reading of results are important.

Biometrics is concerned with the planning of experiments as well as the interpretation of results. A basic concept in biometrics is that there is an infinitely large, hypothetical population of which the experimental group or data are a random sample. The difficulty presented by the inherent variability of biological material is circumvented by estimating the variability and taking it into account when assessing the results.

Experimentation, like other measures employed in research, is not infallible. Inability to demonstrate a supposition experimentally does not prove that it is incorrect.

CHAPTER THREE

CHANCE

"Chance favours only those who know
how to court her."—CHARLES NICOLLE

Illustrations

IT WILL be simpler to discuss the rôle of chance in research if
we first consider some illustrative examples of discoveries in
which it played a part. These anecdotes have been taken from
sources believed to be authentic, and one reference is quoted
for each although in many instances several sources have been
consulted. Only ten are included in this section but seventeen
others illustrating the rôle of chance are to be found in the
Appendix.

Pasteur's researches on fowl cholera were interrupted by the
vacation, and when he resumed he encountered an unexpected
obstacle. Nearly all the cultures had become sterile. He attempted
to revive them by sub-inoculation into broth and injection into
fowls. Most of the sub-cultures failed to grow and the birds
were not affected, so he was about to discard everything and
start afresh when he had the inspiration of re-inoculating the
same fowls with a fresh culture. His colleague Duclaux relates :

"To the surprise of all, and perhaps even of Pasteur, who was
not expecting such success, nearly all these fowls withstood the
inoculation, although fresh fowls succumbed after the usual
incubation period."

This resulted in the recognition of the principle of immunisation
with attenuated pathogens.[31]

The most important method used in staining bacteria is that
discovered by the Danish physician C. Gram. He described how
he discovered the method fortuitously when trying to develop a
double stain for kidney sections. Hoping to stain the nuclei violet
and the tubules brown, he used gentian violet followed by iodine

solution. Gram found that after this treatment the tissue was rapidly decolourised by alcohol but that certain bacteria remained blue-black. The gentian violet and iodine had unexpectedly reacted with each other and with a substance present in some bacteria and not others, thus providing not only a good stain but also a simple test which has proved of the greatest value in distinguishing different bacteria.[109]

While engaged in studying the function of the pancreas in digestion in 1889 at Strasbourg, Professors von Mering and Minkowski removed that organ from a dog by operation. Later a laboratory assistant noticed that swarms of flies were attracted by the urine of the operated dog. He brought this to the attention of Minkowski, who analysed the urine and found sugar in it. It was this finding that led to our understanding of diabetes and its subsequent control by insulin.[22] More recently the Scotsman, Shaw Dunn, was investigating the cause of the kidney damage which follows a severe crush injury to a limb. Among other things he injected alloxan and he found that it caused necrosis of the islet tissue of the pancreas. This unexpected finding has provided a most useful tool in the study of diabetes.[32]

The French physiologist, Charles Richet, was testing an extract of the tentacles of a sea anemone on laboratory animals to determine the toxic dose when he found that a small second dose given some time after the first was often promptly fatal. He was at first so astounded at this result that he could hardly believe that it was due to anything he had done. Indeed he said it was in spite of himself that he discovered induced sensitisation or anaphylaxis and that he would never have believed that it was possible.[22] Another manifestation of the same phenomenon was discovered independently by Sir Henry Dale. He was applying serum to strips of involuntary muscle taken from guinea-pigs when he encountered one that reacted violently to the application of horse serum. Seeking an explanation of this extraordinary observation he found that that guinea-pig had some time previously been injected with horse serum.[27]

It was the usual practice among physiologists to use physiological saline as a perfusion fluid during experiments on isolated frogs' hearts. By this means they could be kept beating for perhaps half an hour. Once at the London University College

Hospital a physiologist was surprised and puzzled to find his frogs' hearts continued to beat for many hours. The only possible explanation he could think of was that it was a seasonal effect and this he actually suggested in a report. Then it was found that the explanation was that his laboratory assistant had used tap water instead of distilled water to make up the saline solution. With this clue it was easy to determine what salts in the tap water were responsible for the increased physiological activity. This was what led Sidney Ringer to develop the solution which bears his name and which has contributed so much to experimental physiology.[27]

Dr. H. E. Durham has left the following written account of the discovery of agglutination of bacteria by antiserum.

"It was a memorable morning in November 1894, when we had all made ready with culture and serum provided by Pfeiffer to test his diagnostic reaction *in vivo*. Professor Gruber called out to me 'Durham! Kommen Sie her, schauen Sie an!' Before making our first injection with the mixtures of serum and vibrios, he had put a specimen under the microscope and there agglutination was displayed. A few days later, we had been making our mixtures in small sterilised glass pots, it happened that none were ready sterilised, so I had to make use of sterile test-tubes; those containing the mixture of culture and serum were left standing for a short time and then I called, 'Herr Professor! Kommen Sie her, schauen Sie an!' the phenomenon of sedimentation was before his eyes! Thus there were two techniques available, the microscopic and the macroscopic."

The discovery was quite unexpected and not anticipated by any hypothesis. It occurred incidentally in the course of another investigation, and macroscopic agglutination was found owing to the fortuitous lack of sterilised glass pots. [I am indebted to Professor H. R. Dean for showing me Durham's manuscript.]

Gowland Hopkins, whom many consider the father of biochemistry, gave his practical class a certain well-known test for proteins to carry out as an exercise, but all the students failed to elicit the reaction. Investigation revealed that the reaction was only obtained when the acetic acid employed contained an impurity, glyoxylic acid, which thereafter became the standard test reagent. Hopkins followed up this clue further and sought the group in the protein with which the glyoxylic acid reacted, and this led him to his famous isolation of tryptophane.[88]

When Weil and Felix were investigating cases of louse-borne typhus in Poland in 1915 they isolated the bacterium known as " Proteus X " from some patients. Thinking it might be the cause of the disease they tried agglutination of the organism with the patients' sera and obtained positive results. It was then found that Proteus X was not the causal organism of the disease; nevertheless agglutination of this organism proved to be a reliable and most valuable means of diagnosing typhus. In the course of their experimental study of this serological reaction Weil and Felix identified the O and H antigens and antibodies, and this discovery in turn opened up a completely new chapter in serology. Later it was found that in Malaya those cases of typhus contracted in the scrub failed to show agglutination to Proteus X19. Strangely enough a new strain of Proteus, obtained from England and believed to be a typical strain of Proteus X19, agglutinated with sera from cases of scrub typhus but not with sera from the cases contracted in the town (shop typhus), which were reacting satisfactorily with the Proteus X19 strain that had been used in many parts of the world. Later it transpired that scrub typhus and shop typhus were two different rickettsial diseases. How it came about that the strain of Proteus sent out from England was not only not typical Proteus X19, but had changed to just what was wanted to diagnose the other disease, remains a profound mystery.[37]

Agglutination of red blood cells of the chick by influenza virus was first observed quite unexpectedly by Hirst and independently by McClelland and Hare when they were examining chick embryos infected with the virus. Fluid containing virus got mixed with blood cells which became agglutinated and the alert and observant scientists quickly followed up this clue. The discovery of this phenomenon has not only revolutionised much of our technique concerned with several viruses, but has opened up a method of approach to fundamental problems of virus-cell relationships.[53, 60] Following this discovery, other workers tried haemagglutination with other viruses and Newcastle disease, fowl plague and vaccinia were found to produce the phenomenon. However it was again by chance observation that haemagglutination with the virus of mumps and later of mouse pneumonia was discovered.

Rickettsiae (microbes closely related to viruses) cause typhus and several other important diseases and are difficult to cultivate. Dr. Herald Cox spent much time and effort trying to improve on methods of growing them in tissue culture and had tried adding all sorts of extracts, vitamins and hormones without achieving anything. One day while setting up his tests he ran short of ·chick embryo tissue for tissue culture, so to make up the balance he used yolk sac which previously he, like everyone else, had discarded. When he later examined these cultures, to his "amazement and surprise", he found terrific numbers of the organisms in those tubes where he had happened to put yolk sac. A few nights later while in bed the idea occurred to him of inoculating the rickettsiae directly into the yolk sac of embryonated eggs. Getting out of bed at 4 a.m. he went to the laboratory and made the first inoculation of rickettsiae into the yolk sac. Thus was discovered an easy way of growing masses of rickettsiae, which has revolutionised the study of the many diseases they cause and made possible the production of effective vaccines against them. [Personal communication.]

Rôle of chance in discovery

These ten examples, together with nineteen others given in the Appendix and some of those in Chapters Four and Eight provide striking illustration of the important part that chance plays in discovery. They are the more remarkable when one thinks of the failures and frustrations usually met in research. Probably the majority of discoveries in biology and medicine have been come upon unexpectedly, or at least had an element of chance in them, especially the most important and revolutionary ones. It is scarcely possible to foresee a discovery that breaks really new ground, because it is often not in accord with current beliefs. Frequently I have heard a colleague, relating some new finding, say almost apologetically, "I came across it by accident." Although it is common knowledge that sometimes chance is a factor in the making of a discovery, the magnitude of its importance is seldom realised and the significance of its rôle does not seem to have been fully appreciated or understood. Books have been written on scientific method omitting any reference to chance or empiricism in discovery.

Perhaps the most striking examples of empirical discoveries are to be found in chemotherapy where nearly all the great discoveries have been made by following a false hypothesis or a so-called chance observation. Elsewhere in this book are described the circumstances in which were discovered the therapeutic effects of quinine, salvarsan, sulphanilamide, diamidine, paraminobenzoic acid and penicillin. Subsequent rational research in each case provided only relatively small improvements. These facts are the more amazing when one thinks of the colossal amount of rational research that has been carried out in chemotherapy.

The research worker should take advantage of this knowledge of the importance of chance in discovery and not pass over it as an oddity or, worse, as something detracting from the credit due to the discoverer and therefore not to be dwelt upon. Although we cannot deliberately evoke that will-o'-the-wisp, chance, we can be on the alert for it, prepare ourselves to recognise it and profit by it when it comes. Merely realising the importance of chance may be of some help to the beginner. We need to train our powers of observation, to cultivate that attitude of mind of being constantly on the look-out for the unexpected and make a habit of examining every clue that chance presents. Discoveries are made by giving attention to the slightest clue. That aspect of the scientist's mind which demands convincing evidence should be reserved for the proof stage of the investigation. In research, an attitude of mind is required for discovery which is different from that required for proof, for discovery and proof are distinct processes. We should not be so obsessed with our hypothesis that we miss or neglect anything not directly bearing on it. With this in mind, Bernard insisted that, although hypotheses are essential in the planning of an experiment, once the experiment is commenced the observer should forget his hypothesis. People who are too fond of their hypotheses, he said, are not well fitted for making discoveries. The anecdote (related in Chapter Eight) about Bernard's work starting from the observation that the rabbits passed clear urine, provides a beautiful example of discovery involving chance, observation and a prepared mind.

A good maxim for the research man is "look out for the unexpected."

It is unwise to speak of luck in research as it may confuse our thinking. There can be no objection to the word when it is used to mean merely chance, but for many people luck is a metaphysical notion which in some mystical way influences events, and no such concept should be allowed to enter into scientific thinking. Nor is chance the only factor involved in these unexpected discoveries, as we shall discuss more fully in the next section. In the anecdotes cited, many of the opportunities might well have been passed over had not the workers been on the look-out for anything that might arise. The successful scientist gives attention to every unexpected happening or observation that chance offers and investigates those that seem to him promising. Sir Henry Dale has aptly spoken of opportunism in this connection. Scientists without the flair for discovery seldom notice or bother with the unexpected and so the occasional opportunity passes without them ever being aware of it. Alan Gregg wrote :

> " One wonders whether the rare ability to be completely attentive to, and to profit by, Nature's slightest deviation from the conduct expected of her is not the secret of the best research minds and one that explains why some men turn to most remarkably good advantage seemingly trivial accidents. Behind such attention lies an unremitting sensitivity." [48]

Writing of Charles Darwin, his son said :

> " Everybody notices as a fact an exception when it is striking and frequent, but he had a special instinct for arresting an exception. A point apparently slight and unconnected with his present work is passed over by many a man almost unconsciously with some half considered explanation, which is in fact no explanation. It was just these things that he seized on to make a start from." [28]

It is of the utmost importance that the rôle of chance be clearly understood. The history of discovery shows that chance plays an important part, but on the other hand it plays only one part even in those discoveries attributed to it. For this reason it is a misleading half-truth to refer to unexpected discoveries as " chance discoveries " or " accidental discoveries ".

If these discoveries were made by chance or accident alone, as many discoveries of this type would be made by any inexperienced scientist starting to dabble in research as by Bernard or Pasteur. The truth of the matter lies in Pasteur's famous saying: "In the field of observation, chance favours only the prepared mind." It is the interpretation of the chance observation which counts. The rôle of chance is merely to provide the opportunity and the scientist has to recognise it and grasp it.

Recognising chance opportunities

In reading of scientific discoveries one is sometimes struck by the simple and apparently easy observations which have given rise to great and far-reaching discoveries making scientists famous. But in retrospect we see the discovery with its significance established. Originally the discovery usually has no intrinsic significance; the discoverer gives it significance by relating it to other knowledge, and perhaps by using it to derive further knowledge. The difficulties in the way of making discoveries in which chance is involved may be discussed under the following headings.

(a) *Infrequency of opportunities.* Opportunities, in the form of significant clues, do not come very often. This is the only aspect affected by sheer chance, and even here the scientist does not play a purely passive rôle. The successful researchers are scientists who spend long hours working at the bench, and who do not confine their activities to the conventional but try out novel procedures, therefore they are exposed to the maximum extent to the risk of encountering a fortunate "accident".

(b) *Noticing the clue.* Acute powers of observation are often required to notice the clue, and especially the ability to remain alert and sensitive for the unexpected while watching for the expected. Noticing is discussed at length in the chapter on observation, and it need only be said here that it is mainly a mental process.

(c) *Interpreting the clue.* To interpret the clue and grasp its possible significance is the most difficult phase of all and requires the "prepared mind". Let us consider some instances of failure to grasp opportunities. The history of discovery teems with instances of lost opportunities—clues noticed but their significance not appreciated. Before Röntgen discovered X-rays, at least one other physicist had noticed evidence of the rays but was merely annoyed. Several people now recall having noticed the inhibition of staphylococcal colonies by moulds before Fleming followed it up to discover penicillin. Scott, for instance, reports that he saw it and considered it only a nuisance and he protests against the view that Fleming's discovery was due to chance, for, he says, it was due mainly to his perspicacity in seizing on the opportunity others had let pass.[83] Another interesting case is related by J. T. Edwards.[33] In 1919 he noticed that one of a group of cultures of Brucella abortus grew much more luxuriantly than the others and that it was contaminated with a mould. He called the attention of Sir John M'Fadyean to this, suggesting it might be of significance, but was greeted with scorn. It was not till later that it was discovered that Br. abortus grew much better in the presence of CO_2, which explains why Edwards' culture had grown much better in the presence of the mould. Bordet and others had casually noticed agglutination of bacteria by antisera, but none had seen the possibilities in it until Gruber and Durham did. Similarly, others had seen the phenomenon of bacteriophage lysis before Twort and D'Herelle. F. M. Burnet for one now admits having seen agglutination of chick embryos' red blood cells in the presence of influenza virus and probably others had too but none followed it up till G. K. Hirst, and McClelland and Hare. Many bacteriologists had seen rough to smooth colony variation in bacteria before Arkwright investigated it and found it to be associated with change in virulence and antigenicity. It is now, of course, one of the fundamental facts in immunology and serology.

Sometimes the significance of the clue which chance brings our way is quite obvious, but at others it is just a trivial incident of significance only for the well prepared mind, the mind loaded with relevant data and ripe for discovery. When the mind has a lot of relevant but loosely connected data and vague ideas, a clarifying idea connecting them up may be helped to crystallise by some small incident. Just as a substance may crystallise out of solution in the presence of a nucleus consisting of a minute crystal with the correct configuration, so did the falling apple provide a model for Newton's mind. Sir Henry Souttar has pointed out that it is the content of the observer's brain, accumulated by years of work, that makes possible the moment of triumph. This aspect of chance observation will be discussed further in the chapters on observation and on intuition.

Anyone with an alertness of mind will encounter during the course of an investigation numerous interesting side issues that might be pursued. It is a physical impossibility to follow up all of these. The majority are not worth following, a few will reward investigation and the occasional one provides the opportunity of a lifetime. How to distinguish the promising clues is the very essence of the art of research. The scientist who has an independent mind and is able to judge the evidence on its merits rather than in light of prevailing conceptions is the one most likely to be able to realise the potentialities in something really new. He also needs imagination and a good fund of knowledge, to know whether or not his observation is new and to enable him to see the possible implications. In deciding whether a line of work should be followed, one should not be put off it merely because the idea has already been thought of by others or even been tried without it leading anywhere. This does not necessarily indicate that it is not good; many of the classic discoveries were anticipated in this way but were not properly developed until the right man came along. Edward Jenner was not the first to inoculate people with cowpox to protect them against smallpox, William Harvey was not the first to postulate circulation of the blood, Darwin was by no means the first to suggest evolution, Columbus was not the first European to go to America, Pasteur was not the first to propound the germ theory of disease, Lister was not the first to use carbolic acid as a wound antiseptic. But these men were the ones who fully developed these ideas and forced them on a reluctant world, and most credit rightly goes to them for bringing the discoveries to fruition. It is not only new ideas that lead to discoveries. Indeed few ideas are entirely original. Usually on close study of the origin of an idea, one finds that others had suggested it or something very like it previously. Charles Nicolle calls these early ideas that are not at first followed up, " precursor ideas ".

36

Exploiting opportunities

When a discovery has passed these hurdles and reached a stage where it is recognised and appreciated by its originator, there are still at least three more ways in which its general acceptance may be delayed.

(d) *Failure to follow up the initial finding.* The initial disclosure may not be made the most of because it may not be followed up and exploited. The most productive scientists have not been satisfied with clearing up the immediate question but having obtained some new knowledge, they made use of it to uncover something further and often of even greater importance. Steinhaeuscr discovered in 1840 that cod-liver oil cured rickets but this enormously important fact remained unproved and no more than an opinion for the next eighty years.[94] In 1903 Theobald Smith discovered that some motile bacilli may exist in culture as the normal motile form or as a non-motile variant, and he demonstrated the significance of these two forms in immunological reactions. This work passed almost unnoticed and was forgotten until the phenomenon was rediscovered in 1917 by Weil and Felix. It is now regarded as one of the fundamental facts in immunological reactions.[91] Fleming described crude preparations of penicillin in 1929, but after a few years he dropped work on it without developing a therapeutic agent. He got no encouragement or assistance from others because they knew of many similar stories that had come to nothing. It was some years later that Florey took the work up from where Fleming left off and developed penicillin as a therapeutic agent.

(e) *Lack of an application.* There may be no possible applications of the discovery until years later. Neufeld discovered a rapid method of typing pneumococci in 1902, but it was not till 1931 that it became of any importance when type-specific serum therapy was introduced. Landsteiner discovered the human blood groups in 1901, but it was not till anticoagulants were found and blood transfusion was developed in the 1914–18 war that Landsteiner's discovery assumed importance and attracted attention.

(f) *Indifference and opposition.* Finally the discovery has to run the gauntlet of scepticism and often resistance on the part of others. This can be one of the most difficult hurdles of all and it is here that the scientist occasionally has to fight and in the past has sometimes even lost his life. The psychology of mental resistance to new ideas, and actual opposition to discoveries are discussed in a later chapter.

Several of the points discussed in this and the preceding section may be illustrated by narrowing the story of Jenner's recognition of the potentialities of vaccination and his exploitation of it. Artificial immunisation against smallpox by means of inoculation with virulent smallpox material (variolation) had long been practised in the Orient. Some say that 1000 years B.C. it was the custom of China to insert material from smallpox lesions into the noses of children, others that variolation was introduced into China from India about A.D. 1000.[12, 75, 108] Variolation was introduced from Constantinople into England about the middle of the eighteenth century and became an accepted though not very popular practice about the time that Edward Jenner was born. When Jenner was serving his apprenticeship between thirteen and eighteen years of age, his attention was called to the local belief in Gloucestershire that people who contracted cow-pox from cattle were subsequently immune to smallpox. Jenner found that the local physicians were mostly familiar with the traditional belief but did not take it seriously, although they also were encountering instances of failure of people to develop infection when given variolation after they had had cow-pox. Jenner evidently kept the matter in mind for years without doing anything about it. After returning to country practice he confided in a friend that he intended trying vaccination. He divulged his intentions under a bond of secrecy because he feared ridicule if they should fail. Meanwhile he was exercising his genius for taking pains and making accurate observation by carrying out experiments in other directions. He was making observations on the temperature and digestion of hibernating animals for John Hunter, experimenting with agricultural fertilisers for Joseph Banks and on his own behalf carrying out studies on how the young cuckoo gets rid of its fellow nestlings. He married at thirty-eight and when his wife had a child he inoculated him with swine-pox and showed he was subsequently immune to smallpox. Still none of his colleagues

—John Hunter among them—took much interest in Jenner's ideas about using cow-pox to vaccinate against smallpox and his first tentative paper on the subject was returned to him and apparently rejected. It was not till he was forty-seven years old (in the memorable year 1796) that he made his first successful vaccination from one human being to another. He transferred material from a pustule on the hand of a milkmaid, Sarah Nelmes, to an eight-year-old boy named James Phipps who thereby gained fame in the same odd way as did Joseph Meister for being the first person to receive Pasteur's treatment for rabies nearly a century later.* This is taken as the classical origin of vaccination but, as is often the case in the history of scientific discovery, the issue is not clear-cut. At least two others had actually performed it earlier but failed to follow it up. Jenner continued his experiments, and in 1798 published his famous *Inquiry,* reporting some twenty-three cases who were either vaccinated or had contracted cow-pox naturally and were subsequently shown to be immune to smallpox. Soon afterwards vaccination was taken up widely and spread throughout the world, despite severe opposition from certain quarters which curiously and interestingly enough persists even to-day in a fairly harmless form. Jenner suffered abuse but honours were soon showered on him from all quarters of the globe.[12, 75]

This history provides an admirable demonstration of how difficult it usually is to recognise the true significance of a new fact. Without knowing the full history one might well suppose Jenner's contribution to medical science a very simple one not meriting the fame subsequently bestowed on it. But neither John Hunter nor any of Jenner's colleagues and contemporaries were able to grasp the potentialities in advance, and similar opportunities had occurred and been let pass in other countries. There was an interval of thirty years after the experimentally minded Jenner himself became interested in the popular belief, before he performed the classical, crucial experiments. With our present conceptions of immunisation and of experimentation this may appear surprising but we must remember how revolutionary the idea was, even given the fact that variolation was an accepted

---

* Meister remained at the Pasteur Institute as concierge until the occupation of Paris by the Germans in 1940, when he committed suicide.

practice. The fact that others who had the same opportunity failed to discover vaccination and that it took Jenner thirty years shows what a difficult discovery it was to make. Animals were at that time regarded with repugnance by most people so the idea of infecting a human being with a disease of animals created utmost disgust. All sorts of dire results were prophesied, including " cow-mania " and " ox-faced children " (one was actually exhibited!). Like many great discoveries it did not require great erudition· and it mainly devolved on having bold-ness and independence of mind to accept a revolutionary idea and imagination to realise its potentialities. But Jenner also had practical difficulties to overcome. He found that cows were subject to various sores on the teats, some of which also affected the milkers but did not give immunity to small-pox. Even present day virus specialists have great difficulty in distinguishing between the different types of sores that occur on cows' teats; and the position is complicated by observations suggesting that an attack of cow-pox does not confer immunity against a second attack of the same disease in the cow, a point Jenner himself noted.

Jenner's discovery has its element of irony which so often lends additional interest to scientific anecdotes. Modern investigators believe that the strains of vaccinia now used throughout the world for many years are not cow-pox but have derived from smallpox. Their origin is obscure but it seems that in the early days cow-pox and smallpox got mixed up and an attenuated strain of smallpox developed and was mistakenly used for cow-pox.

SUMMARY

New knowledge very often has its origin in some quite un-expected observation or chance occurrence arising during an investigation. The importance of this factor in discovery should be fully appreciated and research workers ought deliberately to exploit it. Opportunities come more frequently to active bench workers and people who dabble in novel procedures. Interpreting the clue and realising its possible significance requires knowledge without fixed ideas, imagination, scientific taste, and a habit of contemplating all unexplained observations.

CHAPTER FOUR

HYPOTHESIS

"In science the primary duty of ideas is to be useful and interesting even more than to be 'true'."—WILFRED TROTTER

Illustrations

THE rôle of hypothesis in research can be discussed more effectively if we consider first some examples of discoveries which originated from hypotheses. One of the best illustrations of such a discovery is provided by the story of Christopher Columbus' voyage; it has many of the features of a classic discovery in science. (*a*) He was obsessed with an idea—that since the world is round he could reach the Orient by sailing west, (*b*) the idea was by no means original, but evidently he had obtained some additional evidence from a sailor blown off his course who claimed to have reached land in the west and returned, (*c*) he met great difficulties in getting someone to provide the money to enable him to test his idea as well as in the actual carrying out of the experimental voyage, (*d*) when finally he succeeded he did not find the expected new route, but instead found a whole new world, (*e*) despite all evidence to the contrary he clung to the bitter end to his hypothesis and believed that he had found the route to the Orient, (*f*) he got little credit or reward during his lifetime and neither he nor others realised the full implications of his discovery, (*g*) since his time evidence has been brought forward showing that he was by no means the first European to reach America.

In his early investigations on diphtheria, Löffler showed that in experimental animals dying after inoculation with the diphtheria bacillus, the bacteria remained localised at the site of injection. He suggested that death was caused by toxin produced by the bacteria. Following this hypothesis, Emile Roux made numerous experiments attempting to demonstrate such a toxin in cultures of bacteria, but, try as he might, he could not demonstrate it. However, he persisted in his conviction and finally in desperation he injected the heroic dose of 35 ml. of culture filtrate into a guinea-pig. Rather surprisingly the guinea-pig survived the injection of this volume of fluid and in due course Roux had the satisfaction of seeing the animal die with lesions of diphtheria intoxication. Having established this point Roux was soon able to find out that his difficulties were due to the cultures not having been incubated long enough to produce much toxin, and by prolonged incubation he was able to produce powerfully toxic filtrates. This discovery led to immunisation against diphtheria and the therapeutic use of antiserum.[10]

Following the hypothesis that impulses pass along sympathetic nerves and set up chemical changes producing heat in the skin, Claude Bernard severed the cervical sympathetic nerve in the expectation of it leading to cooling of the rabbit's ear. To his surprise the ear on that side became warmer. He had disconnected the blood vessels of the ear from the nervous influence which normally holds them moderately contracted, resulting in a greater flow of blood and hence warming of the ear. Without at first realising what he had done, he had stumbled on to the fact that the flow of blood through the arteries is controlled by nerves, one of the most important advances in knowledge of circulation since Harvey's classical discovery. An interesting and important illustration of what often happens in the field of observation is provided by Bernard's statement that from 1841 onwards he had repeatedly divided the cervical sympathetic without observing these phenomena which he saw for the first time in 1851. In the previous experiments his attention was directed to the pupil; it was not till he looked for changes in the face and ear that he saw them.[44]

Claude Bernard reasoned that the secretion of sugar by the liver would be controlled by the appropriate nerve, which he supposed was the vagus. Therefore he tried puncturing the origin of the nerve in the floor of the fourth ventricle, and found that the glycogenic function of the liver was greatly increased and the blood sugar rose to such an extent that sugar appeared in the urine. However, Bernard soon realised that, interesting and important as were the results obtained, the hypothesis on which the experiment was founded was quite false because this effect was still obtained even after the vagus had been severed. He again showed his capacity to abandon the original reasoning and followed the new clue. In telling this story he said :

> "We must never be too absorbed by the thought we are pursuing."

This investigation has also interest from another point of view. After his first success in producing diabetes by puncturing the fourth ventricle he had great trouble in repeating it and only succeeded after he had ascertained the exact technique necessary. He was indeed fortunate in succeeding in the first attempt, for otherwise after failing two or three times he would have abandoned the idea.

> "We wish to draw from this experiment another general conclusion . . . negative facts when considered alone never teach us anything. How often must man have been and still must be wrong in this way? It even seems impossible absolutely to avoid this kind of mistake." [15]

Towards the end of the last century nothing was known about the nature and cause of the condition in cows known as milk fever. There was no treatment of any value, and many valuable animals died of it. A veterinarian named Schmidt in Kolding, Denmark, formed an hypothesis that it was an auto-intoxication due to absorption of "colostrum corpuscles and degenerated old epithelial cells" from the udder. So, with the object of "checking the formation of colostral milk and paralysing any existing poison" he treated cases by injecting a solution of potassium iodide into the udder. At first he said that a small amount of air entering the udder during the operation was beneficial because it helped the liberation of free iodine. The treatment was strikingly successful. Later he regarded the injection of copious amounts of air along with the solution as an important part of the treatment, on the ground that the air made it possible to massage the solution into all parts of the udder. The treatment was adopted widely and modified in various ways and soon it was found that the injection of air alone was quite as effective. This treatment based on a false idea became standard practice twenty-five years before the bio-chemical processes involved in milk fever were elucidated; indeed the basic cause of the disease is still not understood, nor do we know why the injection of air usually cures the disease.[81, 82]

An hypothesis may be fruitful, not only for its propounder, but may lead to developments by others. Wassermann himself testified that his discovery of the complement fixation test for syphilis was only made possible by Ehrlich's side-chain theory. Also the development of the Wassermann test has another interesting aspect. Since it was not possible to obtain a culture of the spirochaete which causes syphilis, he used as antigen an extract of liver of syphilitic stillborn children, which he knew contained large numbers of spirochaetes. This worked very well and it was not until some time later that it was found that not only was it unnecessary to use syphilitic liver but equally good antigens could be prepared from normal organs of other animals. To this day it is a mystery why these antigens give a complement fixation reaction which can be used to diagnose syphilis, and only one thing is certain : that the idea that prompted Wassermann to use an extract of liver was entirely fortuitous. But since we still see no reasoned explanation, we would probably still have no serological test for syphilis but for Wasserman's false but fruitful idea.

The foundation of chemotherapy was due to Paul Ehrlich's idea that, since some dyes selectively stained bacteria and protozoa, substances might be found which could be selectively absorbed by the parasites and kill them without damaging the host. His faith in this idea enabled him to persist in the face of long continued frustration, repeated failure and attempts by his friends to dissuade him from the apparently hopeless task. He met with no success until he found that trypan red had some activity against protozoa and, developing further along lines suggested by this, he later developed salvarsan, an arsenical compound effective therapeutically against syphilis, the six hundred and sixth compound of the series. This is perhaps the best example in the history of the study of disease of faith in a hypothesis triumphing over seemingly insuperable difficulties. It would be satisfying to end the story there but, as so often happens, in science, the final note must be one of irony. Ehrlich's search for substances which are selectively absorbed by pathogenic organisms was inspired by his firm belief that drugs cannot act unless fixed to the organisms; but to-day many effective chemotherapeutic drugs are known not to be selectively fixed to the infective agents.

However the story is not yet finished. Gerhard Domagk, impressed by Ehrlich's early work, tried the effects of a great number of dyes belonging to the group called " azo-dyes " to which Ehrlich's trypan red belonged. Then in 1932 he found a dye of this series, prontosil, which was effective therapeutically against streptococci without damaging the infected animal. This discovery marked the beginning of a new era in medicine. But when the French chemist, Trefouël, set to work on the composition of the drug he was amazed to find its action was in no way due to the fact that it was a dye, but was due to it containing sulphanilamide, which is not a dye. Again Ehrlich's false idea had led to a discovery that can justly be described as miraculous. Sulphanilamide had been known to chemists since 1908 but no one had any reason to suspect it had therapeutic properties. It has been said that, had its properties been known, sulphanilamide could have saved 750,000 lives in the 1914–18 war alone.[8] Ehrlich's early work with dyes is said also to be the starting point of the work which led to the discovery of the modern anti-malarial drug atebrin without which the Allies might not have won the war in the Pacific.

Another group of chemotherapeutic substances which were evolved by following an hypothesis is the diamidine group used against the leishmania which causes kala-azar. The idea with which the investigation was started off was to interfere with the natural metabolic processes of the parasite, especially with its glucose metabolism, by using certain derivatives of insulin. One of these, synthalin, was found to have a remarkable leishmanicidal action, but in a dilution far higher than could possibly affect glucose metabolism. Thus, although the hypothesis was wrong, it led to the discovery of a new group of useful drugs.

In certain parts of Great Britain and Western Australia there occurs a nervous disease of sheep known as swayback, the cause of which baffled investigators for years. In Western Australia, H. W. Bennetts for certain reasons suspected that the disease might be due to lead intoxication. To test this hypothesis he treated some sheep with ammonium chloride which is the antidote to lead. The first trial with this gave promising results, which, however, were not borne out by subsequent trials. This suggested that the disease might be due to the deficiency of some mineral which was present in small amounts in the first batch of ammonium chloride. Following up this clue, Bennetts was soon able to show that the disease was due to deficiency of copper, a deficiency never previously known to produce disease in any animal. In Bennetts' own words :

" The solution of the etiology came in Western Australia from an accidental ' lead ' [clue] resulting from the testing of a false hypothesis."[14]

Use of hypothesis in research

Hypothesis is the most important mental technique of the investigator, and its main function is to suggest new experiments or new observations. Indeed, most experiments and many observations are carried out with the deliberate object of testing an hypothesis. Another function is to help one see the significance of an object or event that otherwise would mean nothing. For instance, a mind prepared by the hypothesis of evolution would make many more significant observations on a field excursion than one not so prepared. Hypotheses should be used as tools to uncover new facts rather than as ends in themselves.

The illustrations given above show some of the ways in which hypotheses lead to discoveries. The first thing that arrests attention is the curious and interesting fact that an hypothesis is sometimes very fruitful without being correct—a point that did not escape the attention of Francis Bacon. Several of the illustrations have been selected as striking demonstrations of this point, and it should not be thought that they are a truly representative sample, for correct guesses are more likely to be productive than ones that are wrong, and the fact that the latter are sometimes useful does not detract from the importance of striving for correct explanations. The examples are, however, realistic in that the vast majority of hypotheses prove to be wrong.

When the results of the first experiment or set of observations are in accord with expectations, the experimenter usually still needs to seek further experimental evidence before he can place much confidence in his idea. Even when confirmed by a number of experiments, the hypothesis has been established as true only for the particular circumstances prevailing in the experiments. Sometimes this is all the experimenter claims or requires for he now has a solution of the immediate problem or a working hypothesis on which to plan further investigation of that problem. At other times the value of the hypothesis is as a base from which new lines of investigation branch out in various directions, and it is applied to as many particular cases as possible. If the hypothesis holds good under all circumstances, it may be elevated to the category of a theory or even, if sufficiently profound, a " law ". An hypothesis which is a generalisation cannot, however, be absolutely proved, as is explained in the chapter on Reason; but in practice it is accepted if it has withstood a critical testing, especially if it is in accord with general scientific theory.

When the results of the first experiment or observation fail to support the hypothesis, instead of abandoning it altogether, sometimes the contrary facts are fitted in by a subsidiary clarifying hypothesis. This process of modification may go on till the main hypothesis becomes ridiculously overburdened with *ad hoc* additions. The point at which this stage is reached is largely a matter of personal judgment or taste. The whole edifice is then broken down and supplanted by another that makes a more acceptable synthesis of all the facts now available.

There is an interesting saying that no one believes an hypothesis except its originator but everyone believes an experiment except the experimenter. Most people are ready to believe something based on experiment but the experimenter knows the many little things that could have gone wrong in the experiment. For this reason the discoverer of a new fact seldom feels quite so confident of it as do others. On the other hand other people are usually critical of an hypothesis, whereas the originator identifies himself with it and is liable to become devoted to it. It is as well to remember this when criticising someone's suggestion, because you may offend and discourage him if you scorn the idea. A corollary to this observation that an hypothesis is a very personal matter, is that a scientist usually works much better when pursuing his own than that of someone else. It is the originator who gets both the personal satisfaction and most of the credit if his idea is proved correct, even if he does not do the work himself. A man working on an hypothesis which is not his own often abandons it after one or two unsuccessful attempts because he lacks the strong desire to confirm it which is necessary to drive him to give it a thorough trial and think out all possible ways of varying the conditions of the experiment. Knowing this, the tactful director of research tries to lead the worker himself to suggest the line of work and then lets him feel the idea was his.

Precautions in the use of hypothesis

(a) *Not to cling to ideas proved useless.* Hypothesis is a tool which can cause trouble if not used properly. We must be ready to abandon or modify our hypothesis as soon as it is shown to be inconsistent with the facts. This is not as easy as it sounds. When delighted by the way one's beautiful brain-child seems to explain several otherwise incongruous facts and offers promise of further advances, it is tempting to overlook an observation that does not fit into the pattern woven, or to try to explain it away. It is not at all rare for investigators to adhere to their broken hypotheses, turning a blind eye to contrary evidence, and not altogether unknown for them deliberately to suppress contrary results. If the experimental results or observations are definitely opposed to the hypothesis or if they necessitate unduly complicated or improbable subsidiary hypotheses to accommodate them, one has to discard the idea with as few regrets as possible. It is easier to drop the old hypothesis if one can find a new one to replace it. The feeling of disappointment too will then vanish.

It was characteristic of both Darwin and Bernard that they were ready to drop or modify their hypotheses as soon as they ceased to be supported by the facts observed. The scientist who has a fertile mind and is rich in ideas does not find it so difficult

48

to abandon one found to be unsatisfactory as does the man who has few. It is the latter who is most in danger of wasting time in hanging on to a notion after the facts warrant its discard. Zinsser picturesquely refers to people clinging to sterile ideas as resembling hens sitting on boiled eggs.

On the other hand, faith in the hypothesis and perseverance is sometimes very desirable, as shown by the examples quoted concerning Roux and Ehrlich. Similarly Faraday persisted with his idea in the face of repeated failures before he finally succeeded in producing electric current by means of a magnet. As Bernard observed, negative results mean very little. There is a great difference between (a) stubborn adherence to an idea which is not tenable in face of contrary evidence, and (b) persevering with an hypothesis which is very difficult to demonstrate but against which there is no direct evidence. The investigator must judge the case with ruthless impartiality. However, even when the facts fit into the second category there may come a time when if no progress is being made it is wisest to abandon the attempt, at least temporarily. The hypothesis may be perfectly good but the techniques or knowledge in related fields required for its verification may not yet be available. Sometimes a project is put on one side for years and taken up again when fresh knowledge is available or the scientist has thought of a new approach.

(b) *Intellectual discipline of subordinating ideas to facts.* A danger constantly to be guarded against is that as soon as one formulates an hypothesis, parental affection tends to influence observations, interpretation and judgment; " wishful thinking " is likely to start unconsciously. Claude Bernard said :

" Men who have excessive faith in their theories or ideas are not only ill-prepared for making discoveries; they also make poor observations."

Unless observations and experiments are carried out with safeguards ensuring objectivity, the results may unconsciously be biased. No less an investigator than Gregor Mendel seems to have fallen into this trap, for Fisher[38] has shown that his results were biased in favour of his expectations. The German zoologist, Gatke, was so convinced of the truth of his views on the high speed that birds are capable of that he reported actual

49

observations of birds covering four miles in a minute. He is believed to have been quite sincere but allowed his beliefs to delude him into making false observations.[46]

The best protection against these tendencies is to cultivate an intellectual habit of subordinating one's opinions and wishes to objective evidence and a reverence for things as they really are, and to keep constantly in mind that the hypothesis is only a supposition. As Thomas Huxley so eloquently said:

> " My business is to teach my aspirations to conform themselves to fact, not to try to make facts harmonise with my aspirations. Sit down before fact as a little child, be prepared to give up every preconceived notion, follow humbly wherever nature leads, or you will learn nothing."

An interesting safeguard has been suggested by Chamberlain,[23] namely, the principle of multiple hypotheses in research. His idea was that as many hypotheses as possible should be invented and all kept in mind during the investigation. This state of mind should prompt the observer to look for facts relative to each and may endow otherwise trivial facts with significance. However, I doubt if this method is often practicable. The more usual practice is a succession of hypotheses, selecting the most likely one for trial, and, if it is found wanting, passing on to another.

When Darwin came across data unfavourable to his hypothesis, he made a special note of them because he knew they had a way of slipping out of the memory more readily than the welcome facts.

(c) *Examining ideas critically.* One should not be too ready to embrace a conjecture that comes into the mind; it must be submitted to most careful scrutiny before being accepted even as a tentative hypothesis, for once an opinion has been formed it is more difficult to think of alternatives. The main danger lies in the idea that seems so " obvious " that it is accepted almost without question. It seemed quite reasonable, in cases of cirrhosis of the liver, to rest that organ as much as possible by giving a low protein diet, but recent investigations have shown that this is just what should not be done, for low protein diet can itself cause liver damage. The practice of resting sprained joints was questioned by no one until a few years ago when a bold spirit found they got better much quicker under a regimen of exercise. For many years farmers practised keeping the surface of the soil loose as a mulch, believing this to decrease the loss of water by evaporation. B. A. Keen showed that this belief was based on inadequate experiments and that under most circumstances the practice was useless. He thus saved the community from a great deal of useless expenditure.

(d) *Shunning misconceptions.* Examples have been quoted showing how hypotheses may be fruitful even when wrong, but nevertheless the great majority have to be abandoned as useless. More serious is the fact that false hypotheses or concepts sometimes survive which, far from being productive, are actually responsible for holding up the advance of science. Two examples are the old notion that every metal contains mercury, and the phlogiston doctrine. According to the latter, every combustible substance contains a constituent which is given up on burning, called phlogiston. This notion for long held up the advance of chemistry, and stood in the way of an understanding of combustion, oxidation, reduction, and other processes. It was finally exposed as a fallacy by Lavoisier in 1778, but the great English scientists, Priestley, Watt and Cavendish, clung to the belief for some time afterwards and Priestley had not been converted to the new outlook when he died in 1804.

The exposure of serious fallacies can be as valuable in the advance of science as creative discoveries. Pasteur fought and conquered the notion of spontaneous generation and Hopkins the semi-mystical concept of protoplasm as a giant molecule. Misconceptions in medicine, apart from holding up advances, have been the cause of much harm and unnecessary suffering. For example, the famous Philadelphian physician, Benjamin Rush (1745–1813), gave as an instance of the sort of treatment he meted out :

" From a newly arrived Englishman I took 144 ounces at 12 bleedings in 6 days; four were in 24 hours; I gave within the course of the same 6 days nearly 150 grains of calomel with the usual proportions of jalop and gamboge."[66]

Once ideas have gained credence, they are rarely abandoned

merely because some contrary facts are found. False ideas are only dropped when hypotheses more in accord with the new facts are put forward.

SUMMARY

The hypothesis is the principal intellectual instrument in research. Its function is to indicate new experiments and observations and it therefore sometimes leads to discoveries even when not correct itself.

We must resist the temptation to become too attached to our hypothesis, and strive to judge it objectively and modify or discard it as soon as contrary evidence is brought to light. Vigilance is needed to prevent our observations and interpretations being biased in favour of the hypothesis. Suppositions can be used without being believed.

CHAPTER FIVE

IMAGINATION

" With accurate experiment and observation to work upon, imagination becomes the architect of physical theory."
—TYNDALL

Productive thinking

THIS chapter and the next contain a brief discussion on how ideas originate in the mind and what conditions are favourable for creative mental effort. The critical examination of the processes involved will be rendered easier if I do as I have done in other parts of this book, and make an arbitrary division of what is really a single subject. Consequently much of the material in this chapter should be considered in connection with Intuition and much of the next chapter applies equally to Imagination.

Dewey analyses conscious thinking into the following phases. First comes awareness of some difficulty or problem which provides the stimulus. This is followed by a suggested solution springing into the conscious mind. Only then does reason come into play to examine and reject or accept the idea. If the idea is rejected, our mind reverts to the previous stage and the process is repeated. The important thing to realise is that the conjuring up of the idea is not a deliberate, voluntary act. It is something that happens to us rather than something we do.[29]

In ordinary thinking ideas continually " occur " to us in this fashion to bridge over the steps in reasoning and we are so accustomed to the process that we are hardly aware of it. Usually the new ideas and combinations result from the immediately preceding thought calling up associations that have been developed in the mind by past experience and education. Occasionally, however, there flashes into the mind some strikingly original idea, not based on past associations or at any rate not on associations that are at first apparent. We may suddenly perceive for the first time the connection between several things or ideas, or may take a great leap forward instead of the usual short step where the connections between each pair or set of ideas are well established and " obvious ". These sudden, large progressions occur not only when one is consciously puzzling the problem but also not uncommonly when one is not thinking of anything in particular, or even when one is mildly occupied with something different, and in these circumstances they are often startling. Although there is probably no fundamental difference between these ideas and the less exciting ones that come to us almost continually, and it is not possible to draw any sharp distinction, it will be convenient to consider them separately in the next chapter under the title " intuitions ". In this section we will draw attention to some general features of productive or creative thinking.

Dewey advocates what he calls " reflective thinking ", that is, turning a subject over in the mind and giving it ordered and consecutive consideration, as distinct from the free coursing of ideas through the head. Perhaps the best term for the latter is day-dreaming; it also has its uses, as we shall see presently. But thinking may be reflective and yet be inefficient. The thinker may not be sufficiently critical of ideas as they arise and may be too ready to jump to a conclusion, either through impatience or laziness. Dewey says many people will not tolerate a state of doubt, either because they will not endure the mental discomfort of it or because they regard it as evidence of inferiority.

> " To be genuinely thoughtful, we must be willing to sustain and protract that state of doubt which is the stimulus to thorough enquiry, so as not to accept an idea or make a positive assertion of a belief, until justifying reasons have been found."[29]

Probably the main characteristic of the trained thinker is that he does not jump to conclusions on insufficient evidence as the untrained man is inclined to do.

It is not possible deliberately to create ideas or to control their creation. When a difficulty stimulates the mind, suggested solutions just automatically spring into the consciousness. The variety and quality of the suggestions are functions of how well prepared our mind is by past experience and education pertinent to the particular problem. What we can do deliberately is to prepare our minds in this way, voluntarily direct our thoughts to a certain problem, hold attention on that problem and appraise the various suggestions thrown up by the subconscious mind. The intellectual element in thinking is, Dewey says, what we do with the suggestions after they arise.

Other things being equal, the greater our store of knowledge, the more likely it is that significant combinations will be thrown up. Furthermore, original combinations are more likely to come into being if there is available a breadth of knowledge extending into related or even distant branches of knowledge. As Dr. E. L. Taylor says :

"New associations and fresh ideas are more likely to come out of a varied store of memories and experience than out of a collection that is all of one kind." [90]

Scientists who have made important original contributions have often had wide interests or have taken up the study of a subject different from the one in which they were originally trained. Originality often consists in finding connections or analogies between two or more objects or ideas not previously shown to have any bearing on each other.

In seeking original ideas, it is sometimes useful to abandon the directed, controlled thinking advocated by Dewey and allow one's imagination to wander freely—to day-dream. Harding says all creative thinkers are dreamers. She defines dreaming in these words :

"Dreaming over a subject is simply . . . allowing the will to focus the mind passively on the subject so that it follows the trains of thought as they arise, stopping them only when unprofitable but in general allowing them to form and branch naturally until some useful and interesting results occur." [51]

Max Planck said :

"Again and again the imaginary plan on which one attempts to build up order breaks down and then we must try another. This imaginative vision and faith in the ultimate success are indispensable. The pure rationalist has no place here." [70]

In meditating thus, many people find that visualising the thoughts, forming mental images, stimulates the imagination. It is said that Clerk Maxwell developed the habit of making a mental picture of every problem. Paul Ehrlich was another great advocate of making pictorial representations of ideas, as

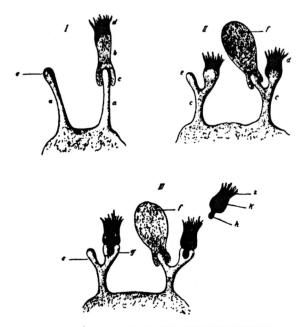

EHRLICH'S DRAWINGS OF HIS SIDE-CHAIN THEORY

one can see from his illustrations of his side-chain theory. Pictorial analogy can play an important part in scientific thinking. This is how the German chemist Kekulé hit on the conception of the benzene ring, an idea that revolutionised organic chemistry. He related how he was sitting writing his chemical text-book:

" But it did not go well; my spirit was with other things. I turned the chair to the fireplace and sank into a half sleep. The atoms flitted before my eyes. Long rows, variously, more closely, united; all in movement wriggling and turning like snakes. And see, what was that? One of the snakes seized its own tail and the image whirled scornfully before my eyes. As though from a flash of lightning I awoke; I occupied the rest of the night in working out the consequences of the hypothesis. . . . Let us learn to dream, gentlemen." [56]

However, physics has reached a stage where it is no longer possible to visualise mechanical analogies representing certain

56

phenomena which can only be expressed in mathematical terms.

In the study of infectious diseases, it is sometimes helpful to take the biological view, as Burnet has done, and look upon the causal organism as a species struggling for continued survival, or even, as Zinsser has felt inclined to do with typhus, which he spent a lifetime studying, personifying the disease in the imagination.

An important inducement to seeking generalisations, especially in physics and mathematics, is the love of order and logical connection between facts. Einstein said :

> " There is no logical way to the discovery of these elemental laws. There is only the way of intuition, which is helped by a feeling for the order lying behind the appearance." [35]

W. H. George remarks that a feeling of tension is produced when an observer sees the objects lying in his field of vision as forming a pattern with a gap in it, and a feeling of relaxation or satisfaction is experienced when the gap is closed, and all parts of the pattern fit into their expected places. Generalisations may be regarded as patterns in ideas.[47] Another phenomenon which may be explained by this concept is the satisfaction experienced on the completion of any task. This may be quite unassociated with any consideration of reward for it applies equally to unimportant, self-appointed tasks such as doing a crossword puzzle, climbing a hill or reading a book. The instinctive sense of irritation we feel when someone disagrees with us or when some fact arises which is contrary to our beliefs may be due to the break in the pattern we have formed.

The tendency of the human mind to seek order in things did not escape the penetrating intelligence of Francis Bacon. He warned against the danger that this trait may mislead us into believing we see a greater degree of order and equality than there really is.

When one has succeeded in hitting upon a new idea, it has to be judged. Reason based on knowledge is usually sufficient in everyday affairs and in straightforward matters in science, but in research there is often insufficient information available for effective reasoning. Here one has to fall back on " feelings " or " taste ". Harding says :

"If the scientist has during the whole of his life observed carefully, trained himself to be on the look out for analogy, and possessed himself of relevant knowledge, then the 'instrument of feeling' . . . will become a powerful divining rod . . . in creative science feeling plays a leading part."[51]

Writing of the importance of imagination in science Tyndall said :

"Newton's passage from a falling apple to a falling moon was an act of the prepared imagination. Out of the facts of chemistry the constructive imagination of Dalton formed the atomic theory. Davy was richly endowed with the imaginative faculty, while with Faraday its exercise was incessant, preceding, accompanying and guiding all his experiments. His strength and fertility as a discoverer are to be referred in great part to the stimulus of the imagination."[95]

Imagination is of great importance not only in leading us to new facts, but also in stimulating us to new efforts, for it enable us to see visions of their possible consequences. Facts and ideas are dead in themselves and it is the imagination that gives life to them. But dreams and speculations are idle fantasies unless reason turns them to useful purpose. Vague ideas captured on flights of fancy have to be reduced to specific propositions and hypotheses.

False trails

While imagination is the source of inspiration in seeking new knowledge, it can also be dangerous if not subjected to discipline; a fertile imagination needs to be balanced by criticism and judgment. This is, of course, quite different from saying it should be repressed or crushed. The imagination merely enables us to wander into the darkness of the unknown where, by the dim light of the knowledge that we carry, we may glimpse something that seems of interest. But when we bring it out and examine it more closely it usually proves to be only trash whose glitter had caught our attention. Things not clearly seen often take on grotesque forms. Imagination is at once the source of all hope and inspiration but also of frustration. To forget this is to court despair.

Most hypotheses prove to be wrong whatever their origin may be. Faraday wrote :

" The world little knows how many of the thoughts and theories which have* passed through the mind of a scientific investigator have been crushed in silence and secrecy by his own severe criticism and adverse examinations; that in the most successful instances not a tenth of the suggestions, the hopes, the wishes, the preliminary conclusions have been realised."

Every experienced research worker will confirm this statement. Darwin went even further :

" I have steadily endeavoured to keep my mind free so as to give up any hypothesis, however much beloved (and I cannot resist forming one on every subject) as soon as facts are shown to be opposed to it. . . . *I cannot remember a single first formed hypothesis which had not after a time to be given up or be greatly modified.*"[28] (Italics mine.)

T. H. Huxley said that the great tragedies of science are the slaying of beautiful hypotheses by ugly facts. F. M. Burnet has told me that most of the " bright ideas " that he gets prove to be wrong.

There is nothing reprehensible about making a mistake, provided it is detected in time and corrected. The scientist who is excessively cautious is not likely to make either errors or discoveries. Whitehead has expressed this aptly : " panic of error is the death of progress." Humphrey Davy said : " The most important of my discoveries have been suggested to me by my failures." The trained thinker shows to great advantage over the untrained person in his reaction to finding his idea to be wrong. The former profits from his mistakes as much as from his successes. Dewey says :

" What merely annoys and discourages a person not accustomed to thinking . . . is a stimulus and guide to the trained enquirer. . . . It either brings to light a new problem or helps to define and clarify the problem."[29]

The productive research worker is usually one who is not afraid to venture and risk going astray, but who makes a rigorous test for error before reporting his findings. This is so not only in the biological sciences but also in mathematics. Hadamard states that good mathematicians often make errors but soon perceive and correct them, and that he himself makes more errors than his students. Commenting on this statement, Sir

Frederic Bartlett, Professor of Psychology at Cambridge, suggests that the best single measure of mental skill may lie in the speed with which errors are detected and thrown out.[11] Lister once remarked:

"Next to the promulgation of the truth, the best thing I can conceive that a man can do is the public recantation of an error."

W. H. George points out that even with men of genius, with whom the birth rate of hypotheses is very high, it only just manages to exceed the death rate.

Max Planck, whose quantum theory is considered by many to be an even more important contribution to science than Einstein's theory of relativity, said when he was awarded the Nobel Prize:

"Looking back . . . over the long and labyrinthine path which finally led to the discovery [of the quantum theory], I am vividly reminded of Goethe's saying that men will always be making mistakes as long as they are striving after something."[70]

Einstein in speaking of the origin of his general theory of relativity said:

"These were errors in thinking which caused me two years of hard work before at last, in 1915, I recognised them as such. . . . The final results appear almost simple; any intelligent undergraduate can understand them without much trouble. But the years of searching in the dark for a truth that one feels, but cannot express; the intense desire and the alternations of confidence and misgiving, until one breaks through to clarity and understanding, are only known to him who has himself experienced them."[35]

Perhaps the most interesting and revealing anecdote on these matters was written by Hermann von Helmholtz[58]:

"In 1891 I have been able to solve a few problems in mathematics and physics including some that the great mathematicians had puzzled over in vain from Euler onwards. . . . But any pride I might have felt in my conclusions was perceptibly lessened by the fact that I knew that the solution of these problems had almost always come to me as the gradual generalisation of favourable examples, by a series of fortunate conjectures, after many errors. I am fain to compare myself with a wanderer on the mountains who, not knowing the path, climbs slowly and painfully upwards and often has to retrace his steps because he can go no further—then, whether by taking thought or from luck, discovers a new track that leads him on a little till at length when he reaches the summit he finds to his shame that there is a royal road, by which he might have ascended, had he only had the wits to find the right approach to it. In my works, I naturally said nothing about my mistake to the reader, but only described the made track by which he may now reach the same heights without difficulty."

Curiosity as an incentive to thinking

In common with other animals we are born with an instinct of curiosity. It provides the incentive for the young to discover the world in which they live—what is hard or soft, movable or fixed, that things fall downwards, that water has the property we call wetness, and all other knowledge required to enable us to accommodate ourselves to our environment. Infants whose mental reflexes have not yet been conditioned are said not to exhibit the " attack-escape " reaction as do adults, but to show rather the opposite type of behaviour. By school age we have usually passed this stage of development, and most of our acquisition of new knowledge is then made by learning from others, either by observing them or being told or reading. We have gained a working knowledge of our environment and our curiosity tends to become blunted unless it is successfully transferred to intellectual interests.

The curiosity of the scientist is usually directed toward seeking an understanding of things or relationships which he notices have no satisfactory explanation. Explanations usually consist in connecting new observations or ideas to accepted facts or ideas. An explanation may be a generalisation which ties together a bundle of data into an orderly whole that can be connected up with current knowledge and beliefs. That strong desire scientists usually have to seek underlying principles in masses of data not obviously related may be regarded as an adult form or sublimation of curiosity. The student attracted to research is usually one who retains more curiosity than usual.

We have seen that the stimulus to the production of ideas is the awareness of a difficulty or problem, which may be the realisation of the present unsatisfactory state of knowledge. People with no curiosity seldom get this stimulus, for one usually

becomes aware of the problem by asking why or how some
process works, or something takes the form that it does. That
a question is a stimulus is demonstrated by the fact that when
someone asks a question it requires an effort to restrain oneself
from responding.

Some purists contend that scientists should wonder "how"
and not "why". They consider that to ask "why" implies that
there is an intelligent purpose behind the design of things and
that activities are directed by a supernatural agency toward
certain aims. This is the teleological view and is rejected by
present-day science, which strives to understand the mechanism
of all natural phenomena. Von Bruecke once remarked:

"Teleology is a lady without whom no biologist can live; yet
he is ashamed to show himself in public with her."

In biology, asking "why" is justified because all events have
causes; and because structures and reactions usually fulfil some
function which has survival value for the organism, and in that
sense they have a purpose. Asking "why" is a useful stimulus
towards imagining what the cause or purpose may be. "How"
is also a useful question in provoking thought about the
mechanism of a process.

There is no satisfying the scientists' curiosity, for with each
advance, as Pavlov said, "we reach a higher level from which
a wider field of vision is open to us, and from which we see
events previously out of range." It may be appropriate to give
here an illustration of how curiosity led John Hunter to carry
out an experiment which led to an important finding.

While in Richmond Park one day Hunter saw a deer with
growing antlers. He wondered what would happen if the blood
supply were shut off on one side of the head. He carried out
the experiment of tying the external carotid artery on one side,
whereupon the corresponding antler lost its warmth and ceased
to grow. But after a while the horn became warm again and
grew. Hunter ascertained that his ligature still held, but neigh-
bouring arteries had increased in size till they carried an adequate
supply of blood. The existence of collateral circulation and the
possibility of its increasing were thus discovered. Hitherto no
one had dared to treat aneurism by ligation for fear of gangrene,
but now Hunter saw the possibilities and tried ligation in the

case of popliteal aneurism. So the Hunterian operation, as it is known in surgery to-day, came into an assured existence.[52] An insatiable curiosity seems to have been the driving force behind Hunter's prolific mind which laid the foundation of modern surgery. He even paid the expenses of a surgeon to go and observe whales for him in the Greenland fisheries.

Discussion as a stimulus to the mind

Productive mental effort is often helped by intellectual intercourse. Discussing a problem with colleagues or with lay persons may be helpful in one of several ways.

(a) The other person may be able to contribute a useful suggestion. It is not often that he can help by directly indicating a solution of the impasse, because he is unlikely to have as much pertinent knowledge as has the scientist working on the problem, but with a different background of knowledge he may see the problem from a different aspect and suggest a new approach. Even a layman is sometimes able to make useful suggestions. For example, the introduction of agar for making solid media for bacteriology was due to a suggestion of the wife of Koch's colleague Hesse.[18]

(b) A new idea may arise from the pooling of information or ideas of two or more persons. Neither of the scientists alone may have the information necessary to draw the inference which can be obtained by a combination of their knowledge.

(c) Discussion provides a valuable means of uncovering errors. Ideas based on false information or questionable reasoning may be corrected by discussion and likewise unjustified enthusiasms may be checked and brought to a timely end. The isolated worker who is unable to talk over his work with colleagues will more often waste his time in following a false trail.

(d) Discussion and exchange of views is usually refreshing, stimulating and encouraging, especially when one is in difficulties and worried.

(e) The most valuable function of discussion is, I believe, to help one to escape from an established habit of thought which has proved fruitless, that is to say, from conditioned thinking. The phenomenon of conditioned thinking is discussed in the next section.

Discussions need to be conducted in a spirit of helpfulness and mutual confidence and one should make a deliberate effort to keep an open receptive mind. Discussions are usually best when not more than about six are present. In such a group no one should be afraid of admitting his ignorance on certain matters and so having it corrected, for in these days of extreme specialisation everyone's knowledge is restricted. Conscious ignorance and intellectual honesty are important attributes for the research man. Free discussion requires an atmosphere unembarrassed by any suggestion of authority or even respect. Brailsford Robertson tells the story of the great biochemist, Jacques Loeb, who, when asked a question by a student after a lecture, replied characteristically :

> " I cannot answer your question, because I have not yet read that chapter in the text-book myself, but if you will come to me to-morrow I shall then have read it, and may be able to answer you." [74]

Students sometimes quite wrongly think that their teachers are almost omniscient, not knowing that the lecturers usually spend a considerable amount of time preparing their lectures, and that outside the topic of the lecture their knowledge is often much less impressive. Not only does the author of a text-book not carry in his head all the information in the book, but the author of a research paper not infrequently has to refer to the paper to recall the details of the work which he himself did.

The custom of having lunch and afternoon tea in groups at the laboratory is a valuable one as it provides ample opportunities for these informal discussions. In addition, slightly more formal seminars or afternoon tea meetings at which workers present their problems before and during, as well as after, the investigation are useful. Sharing of interests and problems among workers in a department or institute is also valuable in promoting a stimulating atmosphere in which to work. Enthusiasm is infectious and is the best safeguard against the doldrums.

Conditioned thinking

Psychologists have observed that once we have made an error, as for example in adding up a column of figures, we have a tendency to repeat it again and again. This phenomenon is known as the persistent error. The same thing happens when we ponder over a problem; each time our thoughts take a certain course, the more likely is that course to be followed the next time. Associations form between the ideas in the chain of thoughts and become firmer each time they are used, until finally the connections are so well established that the chain is very difficult to break. Thinking becomes conditioned just as conditioned reflexes are formed. We may have enough data to arrive at a solution to the problem, but, once we have adopted an unprofitable line of thought, the oftener we pursue it, the harder it is for us to adopt the profitable line. As Nicolle says, " The longer you are in the presence of a difficulty, the less likely you are to solve it."

Thinking also becomes conditioned by learning from others by word of mouth or by reading. In the first chapter we discussed the adverse effect on originality of uncritical reading. Indeed, all learning is conditioning of the mind. Here, however, we are concerned with the effects of conditioning which are unprofitable for our immediate purpose, that of promoting original thought. This does not only concern learning or being conditioned to incorrect opinions for, as we have seen in the first chapter, reading, even the reading of what is true so far as it goes, may have an adverse effect on originality.

The two main ways of freeing our thinking from conditioning are temporary abandonment and discussion. If we abandon a problem for a few days or weeks and then return to it the old thought associations are partly forgotten or less strong and often we can then see it in a fresh light, and new ideas arise. The beneficial effect of temporary abandonment is well shown by laying aside for a few weeks a paper one has written. On coming back to it, flaws are apparent that escaped attention before, and fresh pertinent remarks may spring into the mind.

Discussion is a valuable aid in breaking away from sterile lines of thought that have become fixed. In explaining a problem to another person, and especially to someone not familiar with that field of science, it is necessary to clarify and amplify aspects of it that have been taken for granted and the familiar chain of thought cannot be followed. Not infrequently it happens that while one is making the explanation, a new thought occurs to one without the other person having said a word. The same may happen during the delivery of a lecture, for when the teacher explains something he " sees " it more clearly himself than he had before. The other person, by asking questions, even ill-informed ones, may make the narrator break the established chain, even if only to explain the futility of the suggestion, and this may result in him seeing a new approach to the problem or the connection between two or more observations or ideas that he had not noticed before. The effect that questioning has on the mind might be likened to the stimulus given to a fire by poking; it disturbs the settled arrangement and brings about new combinations. In disturbing fixed lines of thought, discussion is perhaps more likely to be helpful when carried on with someone not familiar with your field of work, for near colleagues have many of the same thought habits as yourself. The writing of a review of the problem may prove helpful in the same way as the giving of a lecture.

A further useful application of the conception of conditioned thinking is that when a problem has defied solution it is best to start again right from the beginning, and if possible with a new approach. For example, I worked unsuccessfully for several years trying to discover the micro-organism which causes foot-rot in sheep. I met with repeated frustrations but each time I started again along the same lines, namely, trying to select the causal organism by microscopy and then isolating it in culture. This method seemed the sensible one to follow and only when I had exhausted all possibilities and was forced to abandon it, did I think of a fundamentally different approach to the problem, namely, to try mixed cultures on various media until one was found which was capable of setting up the disease. Work along these lines soon led to the solution of the problem.

SUMMARY

Productive thinking is started off by awareness of a difficulty. A suggested solution springs into the mind and is accepted or rejected. New combinations in our thoughts arise from rational associations, or from fancy or perhaps chance circumstances. The fertile mind tries a large number and variety of combinations.

66

The scientific thinker becomes accustomed to withholding judgment and remaining in doubt when the evidence is insufficient. Imagination only rarely leads one to a correct answer, and most of our ideas have to be discarded. Research workers ought not to be afraid of making mistakes provided they correct them in good time.

Curiosity atrophies after childhood unless it is transferred to an intellectual plane. The research worker is usually a person whose curiosity is turned toward seeking explanations for phenomena that are not understood.

Discussion is often helpful to productive thinking and informal daily discussion groups in research institutes are valuable.

Once we have contemplated a set of data, the mind tends to follow the same line of thought each time and therefore unprofitable lines of thought tend to be repeated. There are two aids to freeing our thought from this conditioning; to abandon the problem temporarily and to discuss it with another person, preferably someone not familiar with our work.

CHAPTER SIX

INTUITION

"The really valuable factor is intuition."—ALBERT EINSTEIN

Definition and illustration

THE word intuition has several slightly different usages, so it is necessary to indicate at the outset that it is employed here as meaning a sudden enlightenment or comprehension of a situation, a clarifying idea which springs into the consciousness, often, though not necessarily, when one is not consciously thinking of that subject. The terms inspiration, illumination and "hunch" are also used to describe this phenomenon but these words are very often given other meanings. Ideas coming dramatically when one is not consciously thinking of the subject are the most striking examples of intuition, but those arriving suddenly when the problem is being consciously pondered are also intuitions. Usually these were not self-evident when the data were first obtained. All ideas, including the simple ones that form the gradual steps in ordinary reasoning, probably arise by the process of intuition and it is only for convenience that we consider separately in this chapter the more dramatic and important progressions of thought.

Valuable contributions on the subject of intuition in scientific thought have been made by the American chemists Platt and Baker,[71] by the French mathematicians Henri Poincaré[72] and Jacques Hadamard,[50] by W. B. Cannon,[22] the American physiologist, and by Graham Wallas,[99] the psychologist. In writing this chapter I have drawn freely from the excellent article by Platt and Baker who conducted an enquiry on the subject among chemists by questionnaire. The following illustrations are quoted from material collected by them.

"Freeing my mind of all thoughts of the problem I walked briskly down the street, when suddenly at a definite spot which

CLAUDE BERNARD 1813-1878

LOUIS PASTEUR 1822-1895

CHARLES DARWIN 1809-1882

PAUL EHRLICH 1854-1915

THEOBALD SMITH 1859–1934

WALTER B. CANNON 1871–1945

SIR GOWLAND HOPKINS 1861–1947

SIR HENRY DALE 1875–

I could locate to-day—as if from the clear sky above me—an idea popped into my head as emphatically as if a voice had shouted : "

" I decided to abandon the work and all thoughts relative to it, and then, on the following day, when occupied in work of an entirely different type, an idea came to my mind as suddenly as a flash of lightning and it was the solution . . . the utter simplicity made me wonder why I hadn't thought of it before."

" The idea came with such a shock that I remember the exact position quite clearly." [71]

Prince Kropotkin wrote :

" Then followed months of intense thought in order to find out what the bewildering chaos of scattered observations meant until one day all of a sudden the whole became as clear and comprehensible as if it were illuminated with a flash of light . . . There are not many joys in human life equal to the joy of the sudden birth of a generalisation illuminating the mind after a long period of patient research."

Von Helmholtz, the great German physicist said that after previous investigation of a problem " in all directions . . . happy ideas came unexpectedly without effort like an inspiration." He found that ideas did not come to him when his mind was fatigued or when at the working table, but often in the morning after a night's rest or during the slow ascent of wooded hills on a sunny day.

After Darwin had conceived the basic idea of evolution, he was reading Malthus on population for relaxation one day when it struck him that under the struggle for existence favourable variations would tend to be preserved and unfavourable ones destroyed. He wrote a memorandum around this idea, but there was still one important point not accounted for, namely, the tendency in organic beings descended from the same stock to diverge as they become modified. The clarification of this last point came to him under the following circumstances :

" I can remember the very spot in the road, whilst in my carriage, when to my joy the solution occurred to me."

The idea of survival of the fittest as a part of the explanation of evolution also came independently to A. R. Wallace when he was reading Malthus' *Principles of Population* during an illness.

Malthus gave a clear exposition of the checks to increase in the human population and mentioned that these eliminated the least fit. Then it occurred to Wallace that the position was much the same in the animal world.

> "Vaguely thinking over the enormous and constant destruction this implied, it occurred to me to ask the question, 'Why do some die and some live?' and the answer was clearly that on the whole the best fitted live. . . . Then it suddenly flashed upon me that this self-acting process would improve the race . . . the fittest would survive. Then at once I seemed to see the whole effect of this." [89]

Here is Metchnikoff's own account of the origin of the idea of phagocytosis:

> "One day when the whole family had gone to the circus to see some extraordinary performing apes, I remained alone with my microscope, observing the life in the mobile cells of a transparent starfish larva, when a new thought suddenly flashed across my brain. It struck me that similar cells might serve in the defence of the organism against intruders. Feeling that there was in this something of surpassing interest, I felt so excited that I began striding up and down the room and even went to the seashore to collect my thoughts." [62]

Poincaré relates how after a period of intense mathematical work he went for a journey into the country and dismissed his work from mind.

> "Just as I put my foot on the step of the brake, the idea came to me . . . that the transformations I had used to define Fuchsian functions were identical with those of non-Euclidian geometry." [72]

On another occasion when baffled by a problem he went to the seaside and

> "thought of entirely different things. One day, as I was walking on the cliff the idea came to me, again with the same characteristics of conciseness, suddenness and immediate certainty, that arithmetical transformations of indefinite ternary quadratic forms are identical with those of non-Euclidian geometry."

Hadamard cites an experience of the mathematician Gauss, who wrote concerning a problem he had tried unsuccessfully to prove for years, "finally two days ago I succeeded . . . like a sudden flash of lightning the riddle happened to be solved. I cannot myself say what was the conducting thread which connected what I previously knew with what made my success possible."

Intuitions sometimes occur during sleep and a remarkable example is quoted by Cannon. Otto Loewi, professor of pharmacology at the University of Graz, awoke one night with a brilliant idea. He reached for a pencil and paper and jotted down a few notes. On waking next morning he was aware of having had an inspiration during the night, but to his consternation could not decipher his notes. All day at the laboratory in the presence of familiar apparatus he tried to remember the idea and to decipher the note, but in vain. By bedtime he had been unable to recall anything, but during the night to his great joy he again awoke with the same flash of insight. This time he carefully recorded it before going to sleep again.

"The next day he went to his laboratory and in one of the neatest, simplest and most definite experiments in the history of biology brought proof of the chemical mediation of nerve impulses. He prepared two frogs' hearts which were kept beating by means of salt solution. He stimulated the vagus nerve on one of the hearts, thus causing it to stop beating. He then removed the salt solution from this heart and applied it to the other one. To his great satisfaction the solution had the same effect on the second heart as the vagus stimulating had had on the first one: the pulsating muscle was brought to a standstill. This was the beginning of a host of investigations in many countries throughout the world on chemical intermediation, not only between nerves and the muscles and the glands they affect but also between nervous elements themselves." [22]

Cannon states that from his youth he was accustomed to get assistance from sudden and unpredicted insight and that not infrequently he would go to sleep with a problem on his mind and on waking in the morning the solution was at hand. The following passage shows a slightly different use of intuition.

"As a matter of routine I have long trusted unconscious processes to serve me—for example, when I have had to prepare a public address. I would gather points for the address and write them down in rough outline. Within the next few nights I would have sudden spells of awakening, with an onrush of illustrative instances, pertinent phrases, and fresh ideas related to those already listed. Paper and pencil at hand permitted the capture of these fleeting thoughts before they faded into oblivion. The process has been so common and so reliable for me that I have supposed that it was at the service of everyone. But evidence indicates that it is not." [22]

Similarly, in preparing this book ideas have frequently come to me at odd times of the day, sometimes when I was thinking of it, sometimes when I was not. These were all jotted down and later sorted out.

These examples should be ample to enable the reader to understand the particular sense in which I am using the word intuition and to realise its importance in creative thinking.

Most but not all scientists are familiar with the phenomenon of intuition. Among those answering the questionnaire of Platt and Baker 33 per cent reported frequent, 50 per cent occasional, and 17 per cent no assistance from intuition. From other enquiries also it is known that some people, so far as they are aware, never get intuitions, or at any rate not striking ones. They have no comprehension of what an intuition is, and believe that they derive their ideas only from conscious thinking. Some of these opinions may be based on insufficient examination of the working of one's own mind.

The examples cited may leave the reader with the impression that all intuitions are correct or at least fruitful, which, if so, would be inconsistent with what has been said about hypotheses and ideas in general. Unfortunately intuitions, being but the products of fallible human minds, are by no means always correct. In Platt and Baker's enquiry, 7 per cent of scientists replying said their intuitions were always correct, and the remainder gave estimates varying from 10 per cent to 90 per cent of the intuitions as subsequently proving to be correct. Even this is probably an unduly favourable picture, because successful instances would tend to be remembered rather than the unsuccessful. Several eminent scientists have stated that most of their intuitions subsequently prove to be wrong and are forgotten.

Psychology of intuition

The most characteristic circumstances of an intuition are a period of intense work on the problem accompanied by a desire for its solution, abandonment of the work perhaps with attention to something else, then the appearance of the idea with dramatic suddenness and often a sense of certainty. Often there is a feeling of exhilaration and perhaps surprise that the idea had not been thought of previously.

The psychology of the phenomenon is not thoroughly understood. There is a fairly general, though not universal, agreement that intuitions arise from the subconscious activities of the mind which has continued to turn over the problem even though perhaps consciously the mind is no longer giving it attention.

In the previous chapter it was pointed out that ideas spring straight into the conscious mind without our having deliberately formed them. Evidently they originate from the subconscious activities of the mind which, when directed at a problem, immediately brings together various ideas which have been associated with that particular subject before. When a possibly significant combination is found it is presented to the conscious mind for appraisal. Intuitions coming when we are consciously thinking about a problem are merely ideas that are more startling than usual. But some further explanation is needed to account for intuitions coming when our conscious mind is no longer dwelling on that subject. The subconscious mind has probably continued to be occupied with the problem and has suddenly found a significant combination. Now, a new idea arriving during conscious thinking often produces a certain emotional reaction—we feel pleased about it and perhaps somewhat excited. Perhaps the subconscious mind is also capable of reacting in this way and this has the effect of bringing the idea into the conscious mind. This is only a conjecture, but there can be little doubt that a problem may continue to occupy the subconscious mind, for common experience shows that sometimes you "can't get a problem off your mind" because it keeps cropping up involuntarily in your thoughts. Secondly, there is no doubt about the emotion often associated with an intuition.

Some ideas come into consciousness and are grasped, but might not some fail to appear in the conscious mind or only appear fleetingly and disappear again like the things we were about to say but slipped away irretrievably before there was a break in the conversation? According to the hypothesis just outlined the more emotion associated with the idea the more likely it would be to get through to the consciousness. On this reasoning one would expect it to be helpful to have a strong desire for a solution to the problem and also to cultivate a "taste" in scientific matters. It would be interesting to know whether scientists who say they never get intuitions are those who find no joy in new ideas or are deficient in emotional sensitivity.

The conception of the psychology of intuition outlined is in accord with what is known about the conditions that are conducive to their occurrence. It provides an explanation for the importance of (a) freedom from other competing problems and worries, and (b) the helpfulness of periods of relaxation in allowing for the appearance of the intuition, for messages from the subconscious may not be received if the conscious mind is constantly occupied or too fatigued. There have been several instances of famous generalisations coming to people when they were ill in bed. The idea of natural selection in evolution came to Wallace during a bout of malaria, and Einstein has reported that his profound generalisation connecting space and time occurred to him while he was sick in bed. Both Cannon and Poincaré report having got bright ideas when lying in bed unable to sleep—the only good thing to be said for insomnia! It is said that James Brindley, the great engineer, when up against a difficult problem, would go to bed for several days till it was solved. Descartes is said to have made his discoveries while lying in bed in the morning and Cajal refers to those placid hours after awakening which Goethe and so many others considered propitious to discovery. Walter Scott wrote to a friend:

"The half hour between waking and rising has all my life proved propitious to any task which was exercising my invention. . . . It was always when I first opened my eyes that the desired ideas thronged upon me."

Baker finds lying in the bath the ideal time and suggests that Archimedes hit upon his famous principle in the bath because of the favourable conditions and not because he noticed the buoyancy of his body in water. The favourable effects of the bed and the bath are probably due to complete freedom from distraction and to the fact that all the circumstances are conducive to reverie. Others attest to the value of leisure or of relaxing light occupations such as walking in the country or pottering in the garden. Hughlings Jackson used to advise his students to sit in a comfortable chair after the day's work was over and allow their thoughts to wander around things which had interested them during the day and write down the ideas that came.

It is evident that to get bright ideas the scientist needs time for meditation. The favourable effect of temporary abandonment may be to escape from unprofitable conditioned thinking. Intense concentration on a problem too long continued may produce a state of mental blockade such as may occur when you try too hard to recall something that has slipped from your mind.

According to Wallas[99] intuitions always appear at the fringe of consciousness, not at the focus. He considers that an effort should be made to grasp them and that a watch should be kept for valuable ideas in the eddies and backwashes rather than in the main current of thought.

It is said that certain people get some kind of warning preceding an intuition. They become aware that something of that nature is imminent without knowing exactly what it will be. Wallas calls this "intimation". This curious phenomenon does not seem to be at all general.

My colleague, F. M. Burnet, finds that intuitions come to him mainly when he is writing and, unlike most people, rarely when he is relaxing. My own experience is that when I have been concentrating on a subject for several days, it keeps coming back into my mind after I have stopped deliberately working on it. During a lecture, social evening, concert or cinema my thoughts will frequently revert to it and then sometimes after a few moments of conscious thought a new idea will occur. Occasionally the idea springs into the consciousness with little or perhaps no preliminary conscious thinking. The brief preliminary conscious thinking may be similar to Wallas' "intimation", and can easily be missed or forgotten. A number of people have commented on the favourable influence of music but there is by no means universal agreement on this point. I find some, but not all, forms of music conducive to intuitions, both when I am attending

75

an entertainment and when I am writing. The enjoyment of music is rather similar, emotionally, to the enjoyment derived from creative mental activity, and suitable music induces the right mood for productive thought.

Elsewhere mention has been made of the tremendous emotional stimulus many people get when they either make a new discovery or get a brilliant intuition. Possibly this emotional reaction is related to the amount of emotional and mental effort that has been invested, as it were, in the problem. Also there is the sudden release from all the frustrations that have been associated with work on the problem. In this connection it is interesting to note the revealing statement of Claude Bernard:

" Those who do not know the torment of the unknown cannot have the joy of discovery."

Emotional sensitivity is perhaps a valuable attribute for a scientist to possess. In any event the great scientist must be regarded as a creative artist and it is quite false to think of the scientist as a man who merely follows rules of logic and experiment. Some of the masters of the art of research have displayed artistic talents in other directions. Einstein was a keen musician and so was Planck. Pasteur and Bernard early showed considerable promise in painting and play-writing, respectively. Nicolle comments on the interesting and curious fact that the ancient Peruvian language had a single word (*hamavec*) for both poet and inventor.[63]

Technique of seeking and capturing intuitions

It may be useful to recapitulate and set out systematically the conditions which most people find conducive to intuition.

(*a*) The most important prerequisite is prolonged contemplation of the problem and the data until the mind is saturated with it. There must be a great interest in it and desire for its solution. The mind must work consciously on the problem for days in order to get the subconscious mind working on it. Naturally the more relevant data the mind has to work on, the better are the chances of reaching a conclusion.

(*b*) An important condition is freedom from other problems or interests competing for attention, especially worry over private affairs.

Referring to these two prerequisites Platt and Baker say :

" No matter how diligently you apply your conscious thought to your work during office hours, if you are not really wrapped up in your work sufficiently to have your mind unconsciously revert to it at every opportunity, or if you have problems of so much more urgency that they crowd out the scientific problems, then you can expect little in the way of an intuition."

(c) Another favourable condition is freedom from interruption or even fear of interruption or any diverting influence such as interesting conversation within earshot or sudden and excessively loud noises.

(d) Most people find intuitions are more likely to come during a period of apparent idleness and temporary abandonment of the problem following periods of intensive work. Light occupations requiring no mental effort, such as walking in the country, bathing, shaving, travelling to and from work, are said by some to be when intuitions most often appear, probably because under these circumstances there is freedom from distraction or interruption and the conscious mind is not so occupied as to suppress anything interesting arising in the subconscious. Others find lying in bed most favourable and some people deliberately go over the problem before going to sleep and others before rising in the morning. Some find that music has a helpful influence but it is notable that only very few consider that they get any assistance from tobacco, coffee or alcohol. A hopeful attitude of mind may help.

(e) Positive stimulus to mental activity is provided by some form of contact with other minds : (i) discussion with either a colleague or a lay person; (ii) writing a report on the investigation, or giving a talk on it; (iii) reading scientific articles, including those giving views with which one disagrees. When reading articles on topics quite unrelated to the problem, the concept underlying a technique or principle may be absorbed and thrown out again as an intuition relating to one's own work.

(f) Having considered the mental technicalities of deliberately seeking intuitions, there remains one further important practical point. It is a common experience that new ideas often vanish within a minute or so of their appearance if an effort is not made to capture them by focusing attention on them long enough to fix them in the memory. A valuable device which is widely used is to make a habit of carrying pencil and paper and noting down original ideas as they flash into the mind. It is said that Thomas Edison had a habit of jotting down almost every thought that occurred to him, however insignificant it may have appeared at the moment. This technique has also been much used by poets and musicians, and Leonardo da Vinci's notes provide a classical example of its use in the arts. Ideas coming during sleep are likely to be particularly elusive, and some psychologists and scientists always leave a pencil and paper nearby; this is also useful for capturing ideas which occur before one goes to sleep or while lying in bed in the morning. Ideas often make their appearance in the fringe of consciousness when one is reading, writing or otherwise engaged mentally on a theme which it is not desirable to interrupt. These ideas should be roughly jotted down as quickly as possible; this not only preserves them but also serves the useful purpose of getting them "off your mind" with the minimum interruption to the main interest. Concentration requires that the mind should not be distracted by retaining ideas on the fringe of consciousness.

(g) Three very important adverse influences have already been mentioned; interruption, worry and competing interests. It takes time to get your mind "warmed up" and working efficiently on a subject, holding a mass of relevant data on the fringe of consciousness. Interruptions disturb this delicate complex and break the mood. Also mental and physical fatigue, too constant working on the problem (especially under pressure), petty irritations and really distracting types of noise can militate against creative thinking. These remarks do not conflict with what is said in Chapter Eleven about the best work sometimes being done under adversity and mental stress. There I am referring rather to the deep-seated problems of life which sometimes may drive one to work in an attempt to escape them. In this chapter I am speaking of the immediate problems of everyday life.

Scientific taste

This seems the most appropriate place to discuss the concept "scientific taste". Hadamard and others have made the interesting observation that there is such a thing as scientific taste, just as there is a literary and an artistic taste.[50] Dale speaks of "the subconscious reasoning which we call instinctive judgment",[27] W. Ostwald[67] refers to "scientific instinct", and some people use the words "intuition" and "feeling" in this connection, by which they mean the same thing, but it seems to me more correct to call this faculty taste. It is probably synonymous with "personal judgment", which some scientists would probably prefer, but I think that expression is even less illuminating than is "taste". It is perhaps more exact to say that taste is that on which we base our personal judgment.

Taste can perhaps best be described as a sense of beauty or aesthetic sensibility, and it may be reliable or not, depending on the individual. Anyone who has it simply feels in his mind that a particular line of work is of interest for its own sake and worth following, perhaps without knowing why. How reliable one's feelings are can be determined only by the results. The concept of scientific taste may be explained in another way by saying that the person who possesses the flair for choosing profitable lines of investigation is able to see further whither the work is leading than are other people, because he has the habit of using his imagination to look far ahead instead of restricting his thinking to established knowledge and the immediate problem. He may not be able to state explicitly his reasons or envisage any particular hypothesis, for he may see only vague hints that it leads towards one or another of several crucial questions.

An illustration of taste in non-scientific matters is the choice of words and composition of sentences when writing. Only occasionally is it necessary to check the correctness of the language used by submitting it to grammatical analysis; usually we just "feel" that the sentence is correct or not. The elegance and aptness of the English which is produced largely automatically is a function of the taste we have acquired by training in choice and arrangement of words. In research, taste plays an important part in choosing profitable subjects for investigation, in recognising promising clues, in intuition, in deciding on a course of action where there are few facts with which to reason, in discarding hypotheses that require too many modifications and in forming an opinion on new discoveries before the evidence is decisive.

Although, as with other tastes, people may be endowed with the capacity for scientific taste to varying degrees, it may also be cultivated by training oneself in the appreciation of science, as, for example, in reading about how discoveries have been made. As with other tastes, taste in science will only be found in people with a genuine love of science. Our taste derives from the summation of all that we have learnt from others, experienced and thought.

Some scientists may have difficulty in comprehending such an abstract concept as taste, and some may find it unacceptable, because all the scientist's training is toward making him eliminate subjective influences from his work. No one would dispute the policy of keeping the subjective element out of experimentation, observation and technical procedures to the greatest possible extent. How far such a policy can effectively be carried out in a scientist's thinking is more open to question. Most people do not realise how often opinions that are supposed to be based on reason are in fact but rationalisations of prejudice or subjective motives. There is a very considerable part of scientific thinking where there is not enough sound knowledge to allow of effective reasoning and here the judgment will inevitably be largely influenced by taste. In research we continually have to take action on issues about which there is very little direct evidence. Therefore, rather than delude ourselves, I think it is wise to face the fact of subjective judgment and accept the concept of scientific taste, which seems a useful one. But by accepting the idea, I do not mean to suggest that we should adopt taste as a guide in cases where there is enough evidence on which to base an objectively reasoned judgment. The phrase, "scientific taste", must not be allowed to blind us to the risks which are associated with all subjective thinking.

SUMMARY

Intuition is used here to mean a clarifying idea that springs suddenly into the mind. It by no means always proves to be correct.

The conditions most conducive to intuitions are as follows : (a) The mind must first be prepared by prolonged conscious puzzling over the problem. (b) Competing interests or worries are inimical to intuitions. (*c*) Most people require freedom from interruptions and distractions. (*d*) Intuitions often make their appearance when the problem is not being worked on. (*e*) Positive stimuli are provided by intellectual contacts with other minds such as in discussion, critical reading or writing. (*f*) Intuitions often disappear from the mind irretrievably as quickly as they come, so should be written down. (*g*) Unfavourable influences include, in addition to interruptions, worry and competing interests, also mental or physical fatigue, too constant working on a problem, petty irritations and distracting types of noises.

Often in research our thoughts and actions have to be guided by personal judgment based on scientific taste.

CHAPTER SEVEN

REASON

"Discovery should come as an adventure rather than as the result of a logical process of thought. Sharp, prolonged thinking is necessary that we may keep on the chosen road, but it does not necessarily lead to discovery."
—THEOBALD SMITH

Limitations and hazards

BEFORE considering the rôle of reason in research it may be useful to discuss the limitations of reason. These are more serious than most people realise, because our conception of science has been given us by teachers and authors who have presented science in logical arrangement and that is seldom the way in which knowledge is actually acquired.

Everyday experience and history teach us that in the biological and medical sciences reason seldom can progress far from the facts without going astray. The scholasticism and authoritarianism prevailing during the Middle Ages was incompatible with science. With the Renaissance came a change in outlook : the belief that things ought and must behave according to accepted views (mostly taken from the classics) was supplanted by a desire to observe things as they really are, and human knowledge began to grow again. Francis Bacon had a great influence on the development of science mainly, I think, because he showed that most discoveries had been made empirically rather than by use of deductive logic. In 1605 he said :

"Men are rather beholden ... generally to chance, or anything else, than to logic, for the invention of arts and sciences ",[6]

and in 1620,

"the present system of logic rather assists in confirming and rendering inveterate the errors founded on vulgar notions, than in searching after truth, and is therefore more hurtful than useful."[7]

Later the French philosopher René Descartes made people realise that reason can land us in endless fallacies. His golden rule was:

" Give unqualified assent to no propositions but those the truth of which is so clear and distinct that they cannot be doubted."

Every child, indeed one might even say, every young vertebrate, discovers gravity; and yet modern science with all its knowledge cannot yet satisfactorily "explain" it. Not only are reason and logic therefore insufficient to provide a means of discovering gravity without empirical knowledge of it, but all the reason and logic applied in classical times did not even enable intelligent men to deduce correctly the elementary facts concerning it.

F. C. S. Schiller, a modern philosopher, has made some illuminating comments on the use of logic in science and I shall quote from him at length:

" Among the obstacles to scientific progress a high place must certainly be assigned to the analysis of scientific procedure which logic has provided. . . . It has not tried to describe the methods by which the sciences have actually advanced, and to extract . . . rules which might be used to regulate scientific progress, but has freely re-arranged the actual procedure in accordance with its prejudices, for the order of discovery there has been substituted an order of proof." [80]

Credence of the logician's view has been encouraged by the method generally adopted in the writing of scientific papers. The logical presentation of results which is usually followed is hardly ever a chronological or full account of how the investigation was actually carried out, for such would often be dull and difficult to follow and, for ordinary purposes, wasteful of space. In his book on the writing of scientific papers, Allbutt specifically advocates that the course of the research should not be followed but that a deductive presentation should be adopted.

To quote again from Schiller, who takes an extreme view:

" It is not too much to say that the more deference men of science have paid to logic, the worse it has been for the scientific value of their reasoning. . . . Fortunately for the world, however, the great men of science have usually been kept in salutary ignorance of the logical tradition." [80]

He goes on to say that logic was developed to regulate debates in the Greek schools, assemblies and law-courts. It was necessary to determine which side won, and logic served this purpose, but it should not occasion surprise that it is quite unsuitable in science, for which it was never intended. Many logicians emphatically declare that logic, interested in correctness and validity, has nothing at all to do with productive thinking.

Schiller goes even further in his criticism of traditional logic and says that not only is it of little value in making new discoveries, but that history has shown it to be of little value in recognising their validity or ensuring their acceptance when they have been proclaimed. Indeed, logical reasoning has often prevented the acceptance of new truths, as is illustrated by the persecution to which the great discoverers have so often been subjected.

" The slowness and difficulty with which the human race makes discoveries and its blindness to the most obvious facts, if it happens to be unprepared or unwilling to see them, should suffice to show that there is something gravely wrong about the logician's account of discovery."

Schiller was protesting mainly against the view of the scientific method expounded by certain logicians in the latter half of the nineteenth century. Most modern philosophers concerning themselves with the scientific method do not interpret this phrase as including the art of discovery, which they consider to be outside their province. They are interested in the philosophical implications of science.

Wilfred Trotter[94] also had some provocative things to say about the poor record which reason has in the advancement of scientific knowledge. Not only has it few discoveries to its credit compared to empiricism, he says, but often reason has obstructed the advance of science owing to false doctrines based on it. In medicine particularly, practices founded on reason alone have often prevailed for years or centuries before someone with an independent mind questioned them and in many cases showed they were more harmful than beneficial.

Logicians distinguish between inductive reasoning (from particular instances to general principles, from facts to theories) and deductive reasoning (from the general to the particular, applying a theory to a particular case). In induction one starts from observed data and develops a generalisation which explains the relationships between the objects observed. On the other hand, in deductive reasoning one starts from some general law and applies it to a particular instance. Thus in deductive reasoning the derived conclusion is contained within the original premiss, and should be true if the premiss is true.

Since deduction consists of applying general principles to further instances, it cannot lead us to new generalisations and so cannot give rise to major advances in science. On the other hand the inductive process is at the same time less trustworthy but more productive. It is more productive because it is a means of arriving at new theories, but is less trustworthy because starting from a collection of facts we can often infer several possible theories, all of which cannot be true as some may be mutually incompatible; indeed none of them may be true.

In biology every phenomenon and circumstance is so complex and so poorly understood that premisses are not clear-cut and hence reasoning is unreliable. Nature is often too subtle for our reasoning. In mathematics, physics and chemistry the basic premisses are more firmly established and the attendant circumstances can be more rigidly defined and controlled. Therefore reason plays a rather more dominant part in extending knowledge in these sciences. Nevertheless the mathematician Poincaré said: "Logic has very little to do with discovery or invention." Similar views were expressed by Planck and Einstein (pp. 55, 57). The point here is that inductions are usually arrived at not by the mechanical application of logic but by intuition, and the course of our thoughts is constantly guided by our personal judgment. On the other hand the logician is not concerned with the way the mind functions but with logical formulation.

From his experience in finding that his hypotheses always had to be abandoned or at least greatly modified Darwin learnt to distrust deductive reasoning in the biological sciences. He said:

"I must begin with a good body of facts, and not from principle, in which I always suspect some fallacy."[28]

A basic difficulty in applying reason in research derives from the fact that terms often cannot be defined accurately and premisses are seldom precise or unconditionally true. Especially in biology premisses are only true under certain circumstances. For careful reasoning and clarity of thought one should first define the terms one uses but in biology exact definitions are often difficult or impossible to arrive at. Take, for example, the statement "influenza is caused by a virus." Influenza was originally a clinical concept, that is to say, a disease defined on clinical characters. We now know that diseases caused by several different microbes have been embraced by what the clinician regards as influenza. The virus worker would now prefer to define influenza as a disease caused by a virus with certain characters. But this only passes on the difficulty to the defining of an influenza virus which in turn escapes precise definition.

These difficulties are to some extent resolved if we accept the principle that in all our reasoning we can deal only in probabilities. Indeed much of our reasoning in biology is more aptly termed speculation.

I have mentioned some limitations inherent in the application of logical processes in science; another common source of error is incorrect reasoning, such as committing some logical fallacy. It is a delusion that the use of reason is easy and needs no training or special caution. In the following section I have tried to outline some general precautions which it may be helpful to keep in mind in using reason in research.

Some safeguards in use of reason in research

The first consideration is to examine the basis from which we start reasoning. This involves arriving at as clear an understanding as possible of what we mean by the terms we employ, and examining our premisses. Some of the premisses may be well-established facts or laws, while others may be purely suppositions. It is often necessary to admit provisionally some assumptions that are not well established, in which case one needs to be careful not to forget that they are only suppositions. Michael Faraday warned against the tendency of the mind " to rest on an assumption " and when it appears to fit in with other knowledge to forget that it has not been proved. It is generally agreed that unverified assumptions should be kept down to the bare minimum and the hypothesis with the fewest assumptions is to be preferred. (This is known as the maxim of parsimony, or "Occam's Razor". It was propounded by William of Occam in the fourteenth century.)

How easy it is for unverified assumptions to creep into our reasoning unnoticed! They are often introduced by expressions such as "obviously", "of course", "surely". I would have thought that it was a fairly safe assumption that well-fed animals live longer on the average that underfed ones, but in recent experiments mice whose diet was restricted to a point where their growth rate was below normal lived much longer than mice allowed to eat as much as they wished.

Having arrived at a clear understanding of the basis from which we start, at every step in our reasoning it is essential to pause and consider whether all conceivable alternatives have been taken into account. The degree of uncertainty or supposition is usually greatly magnified at each step.

It is important not to confuse facts with their interpretations, that is to say, to distinguish between data and generalisations. Facts are particular observational data relating to the past or present. To take an obvious illustration : it may be a fact that when a certain drug was administered to rabbits it killed them, but to say that the drug is poisonous for rabbits is not a statement of a fact but a generalisation or law arrived at by induction. The change from the past tense to the present usually involves stepping from the facts to the induction. It is a step which must often be taken but only with an understanding of what one is doing. Confusion may also arise from the way in which the results are interpreted : strictly the facts arising from experiments can only be described by a precise statement of what occurred. Often in describing an experiment we interpret the results into other terms, perhaps without realising we are departing from a statement of the facts.

A difficulty we are always up against is that we have to argue from past and present to the future. Science, to be of value, must predict. We have to reason from data obtained in the past by experiment and observation, and plan accordingly for the future. This presents special difficulties in biology because, owing to the incompleteness of our knowledge, we can seldom be sure that changed circumstances in the future may not influence the results.

Take, for example, the testing of a new vaccine against a disease. The vaccine may prove effective in several experiments but we must still be cautious in saying it will be effective in future. Influenza vaccine gave a considerable degree of protection in large scale trials in U.S.A. in 1943 and 1945, but against the next epidemic in 1947 it was of no value. Regarded as a problem in logic the position is that by inductive inference from our data we arrive at a generalisation (for instance, that the vaccine is effective). Then in future when we wish to guard against the disease we use this generalisation deductively and apply it to the particular practical problem of protecting certain people. The difficult point in the reasoning is, of course, making the induction. Logic has little to say here that is of help to us. All we can do is to refrain from generalising until we have collected fairly extensive data to provide a wide basis for the induction and regard as tentative any conclusion based on induction or, as we more often hear in everyday language, be cautious with generalisations. Statistics help us in drawing conclusions from our data by ensuring that our conclusions have a certain reliability, but even statistical conclusions are strictly valid only for events which have already occurred.

Generalisations can never be *proved*. They can be tested by seeing whether deductions made from them are in accord with experimental and observational facts, and if the results are not as predicted, the hypothesis or generalisation may be *disproved*. But a favourable result does not prove the generalisation, because the deduction made from it may be true without its being true. Deductions, themselves correct, may be made from palpably absurd generalisations. For instance, the truth of the hypothesis that plague is due to evil spirits is not established by the correctness of the deduction that you can avoid the disease by keeping out of the reach of the evil spirits. In strict logic a generalisation is never proved and remains on probation indefinitely, but if it survives all attempts at disproof it is accepted in practice, especially if it fits well into a wider theoretical scheme.

If scientific logic shows we must be cautious in arriving at generalisations ourselves, it shows for the same reasons that we should not place excessive trust in any generalisation, even widely accepted theories or laws. Newton did not regard the laws he formulated as the ultimate truth, but probably most following him did until Einstein showed how well-founded Newton's caution had been. In less fundamental matters how often do we see widely accepted notions superseded!

Therefore the scientist cannot afford to allow his mind to become fixed, with reference not only to his own opinions but also to prevailing ideas. Theobald Smith said:

> "Research is fundamentally a state of mind involving continual re-examination of doctrines and axioms upon which current thought and action are based. It is, therefore, critical of existing practices."[85]

No accepted idea or "established principle" should be regarded as beyond being questioned if there is an observation challenging it. Bernard wrote:

> "If an idea presents itself to us, we must not reject it simply because it does not agree with the logical deductions of a reigning theory."

Great discoveries have been made by means of experiments devised with complete disregard for well accepted beliefs. Evidently it was Darwin who introduced the expression "fool's experiment" to refer to such experiments, which he often undertook to test what most people would consider not worth testing.

People in most other walks of life can allow themselves the indulgence of fixed ideas and prejudices which make thinking so much easier, and for all of us it is a practical necessity to hold definite opinions on many issues in everyday life, but the research worker must try to keep his mind malleable and avoid holding set ideas in science. We have to strive to keep our mind receptive and to examine suggestions made by others fairly and on their own merits, seeking arguments for as well as against them. We must be critical, certainly, but beware lest ideas be rejected because an automatic reaction causes us to see only the arguments against them. We tend especially to resist ideas competing with our own.

A useful habit for scientists to develop is that of not trusting ideas based on reason only. As Trotter says, they come into the mind often with a disarming air of obviousness and certainty. Some consider that there is no such thing as pure reasoning, that is to say, except where mathematical symbols are involved.

89

Practically all reasoning is influenced by feelings, prejudice and past experience, albeit often subconsciously. Trotter wrote :

> " The dispassionate intellect, the open mind, the unprejudiced observer, exist in an exact sense only in a sort of intellectualist folk-lore; states even approaching them cannot be reached without a moral and emotional effort most of us cannot or will not make."

A trick of the mind well known to psychologists is to " rationalise ", that is, to justify by reasoned argument a view which in reality is determined by preconceived judgment in the subconscious mind, the latter being governed by self-interest, emotional considerations, instinct, prejudice and similar factors which the person usually does not realise or admit even to himself. In somewhat similar vein is W. H. George's warning against believing that things in nature ought to conform to certain patterns or standards and regarding all exceptions as abnormal. He says that the " should-ought mechanism " has no place whatever in research, and its complete abandonment is one of the foundation stones of science. It is premature, he considers, to worry about the technique of experimentation until a man has become dissatisfied with the " should-ought " way of thinking.

It has been said by some that scientists should train themselves to adopt a disinterested attitude to their work. I cannot agree with this view and think the investigator should try to exercise sufficient self-control to consider fairly the evidence against a certain outcome for which he fervently hopes, rather than to try to be disinterested. It is better to recognise and face the danger that our reasoning may be influenced by our wishes. Also it is unwise to deny ourselves the pleasure of associating ourselves whole-heartedly with our ideas, for to do so would be to undermine one of the chief incentives in science.

It is important to distinguish between interpolation and extrapolation. Interpolating means filling in a gap *between* established facts which form a series. When one draws a curve on a graph by connecting the points one interpolates. Extrapolating is going *beyond* a series of observations on the assumption that the same trend continues. Interpolation is considered permissible for most purposes provided one has a good series of data to work from, but extrapolation is much more hazardous. Apparently obvious extensions of our theories beyond the field in which they have been tested often lead us astray. The process of extrapolation is rather similar to implication and is useful in providing suggestions.

A useful aid in getting a clear understanding of a problem is to write a report on all the information available. This is helpful when one is starting on an investigation, when up against a difficulty, or when the investigation is nearing completion. Also at the beginning of an investigation it is useful to set out clearly the questions for which an answer is being sought. Stating the problem precisely sometimes takes one a long way toward the solution. The systematic arrangement of the data often discloses flaws in the reasoning, or alternative lines of thought which had been missed. Assumptions and conclusions at first accepted as " obvious " may even prove indefensible when set down clearly and examined critically. Some institutions make it a rule for all research workers to furnish a report quarterly on the work done, and work planned. This is useful not only for the director to keep in touch with developments but also to the workers themselves. Certain directors prefer verbal reports which they consider more useful in helping the research worker " get his ideas straight ".

Careful and correct use of language is a powerful aid to straight thinking, for putting into words precisely what we mean necessitates getting our own minds quite clear on what we mean. It is with words that we do our reasoning, and writing is the expression of our thinking. Discipline and training in writing is probably the best training there is in reasoning. Allbutt has said that slovenly writing reflects slovenly thinking, and obscure writing usually confused thinking. The main aim in scientific reports is to be as clear and precise as possible and make each sentence mean exactly what it is intended to and be incapable of other interpretation. Words or phrases that do not have an exact meaning are to be avoided because once one has given a name to something, one immediately has a feeling that the position has been clarified, whereas often the contrary is true. " A verbal cloak of ignorance is a garment that often hinders progress." [91]

The rôle of reason in research

Although discoveries originate more often from unexpected experimental results or observations, or from intuitions, than directly from logical thought, reason is the principle agent in most other aspects of research and the guide to most of our actions. It is the main tool in formulating hypotheses, in judging the correctness of ideas conjured up by imagination and intuition, in planning experiments and deciding what observations to make, in assessing the evidence and interpreting new facts, in making generalisations and finally in finding extensions and applications of a discovery.

The methods and functions of discovery and proof in research are as different as are those of a detective and of a judge in a court of law. While playing the part of the detective the investigator follows clues, but having captured his alleged fact, he turns judge and examines the case by means of logically arranged evidence. Both functions are equally essential but they are different.

It is in "factual" discoveries in biology that observation and chance—empiricism—plays such an important part. But facts obtained by observation or experiment usually only gain significance when we use reason to build them into the general body of knowledge. Darwin said:

"Science consists in grouping facts so that general laws or conclusions may be drawn from them."[28]

In research it is not sufficient to collect facts; by interpreting them, by seeing their significance and consequences we can often go much further. Walshe considers that just as important as making discoveries is what we make *of* our discoveries, or for that matter, of those of other people.[100] To help retain and use information our minds require a rationalised, logically consistent body of knowledge. Hughlings Jackson said that

"We have multitudes of facts, but we require, as they accumulate, organisations of them into higher knowledge; we require generalisations and working hypotheses."

The recognition of a new general principle is the consummation of scientific study.

Discoveries originating from so-called chance observations, from unexpected results in experiments or from intuitions are dramatic and arrest attention more than progress resulting from purely rational experimentation in which each step follows logically on the previous one so that the discovery only gradually unfolds. Therefore the latter, less spectacular process may be responsible for more advances than has been implied in the other chapters of this book. Moreover, as Zinsser said :

> " The preparatory accumulation of minor discoveries and of accurately observed details . . . is almost as important for the mobilisation of great forward drives as the periodic correlation of these disconnected observations into principles and laws by the vision of genius." [108]

Often when one looks into the origin of a discovery one finds that it was a much more gradual process than one had imagined.

In nutritional research, the discovery of the existence of the various vitamins was in a number of instances empirical, but subsequent development of knowledge of them was rational. Usually in chemotherapy, after the initial empirical discovery opening up the field, rational experimentation has led to a series of improvements, as in the development of sulphathiazole, sulphamerazine, sulphaguanidine, etc., following on the discovery of the therapeutic value of sulphanilamide, the first compound of this type found to have bacteriostatic properties.

As described in the Appendix, Fleming followed up a chance observation to discover that the mould *Penicillium notatum* produced a substance that had bacteriostatic properties and was non-toxic. However, he did not pursue it sufficiently to develop a chemotherapeutic agent and the investigation was dropped. During the latter quarter of the last century and first part of this there were literally dozens of reports of discoveries of antibacterial substances produced by bacteria and fungi.[43] Even penicillin itself was discovered before Fleming or Florey.[114] Quite a number of writers had not only suggested that these products might be useful therapeutically but had employed them and in some instances good results seem to have been obtained.[43] But all these empirical discoveries were of little consequence until Florey, by a deliberately planned, systematic attack on the problem, produced penicillin in a relatively pure and stable form and so was able to demonstrate its great clinical value. Often the original discovery, like the crude ore from the mine, is of little value until it has been refined and fully developed. This latter process, less spectacular and largely rational, usually requires a different type of scientist and often a team. The rôle of reason in research is not so much in exploring the frontiers of knowledge as in developing the findings of the explorers.

A type of reasoning not yet mentioned is reasoning by analogy, which plays an important part in scientific thought. An analogy is a resemblance between the relationship of things, rather than between the things themselves. When one perceives that the relationship between A and B resembles the relationship between X and Y on one point, and one knows that A is related to B in various other ways, this suggests looking for similar relationships between X and Y. Analogy is very valuable in suggesting clues or hypotheses and in helping us comprehend phenomena and occurrences we cannot see. It is continually used in scientific thought and language but it is as well to keep in mind that analogy can often be quite misleading and of course can never prove anything.

Perhaps it is relevant to mention here that modern scientific philosophers try to avoid the notion of cause and effect. The current attitude is that scientific theories aim at describing associations between events without attempting to explain the relationship as being causal. The idea of cause, as implying an inherent necessity, raises philosophical difficulties and in theoretical physics the idea can be abandoned with advantage as there is then no longer the need to postulate a connection between the cause and effect. Thus, in this view, science confines itself to description— " how ", not " why ".

This outlook has been developed especially in relation to theoretical physics. In biology the concept of cause and effect is still used in practice, but when we speak of *the* cause of an event we are really over-simplifying a complex situation. Very many factors are involved in bringing about an event but in practice we commonly ignore or take for granted those that are always present or well-known and single out as *the* cause one factor which is unusual or which attracts our attention for a special reason. The cause of an outbreak of plague may be regarded by the bacteriologist as the microbe he finds in the blood of the victims, by the entomologist as the microbe-carrying fleas that spread the disease, by the epidemiologist as the rats that escaped from the ship and brought the infection into the port.

94

SUMMARY

The origin of discoveries is beyond the reach of reason. The rôle of reason in research is not hitting on discoveries—either factual or theoretical—but verifying, interpreting and developing them and building a general theoretical scheme. Most biological " facts " and theories are only true under certain conditions and our knowledge is so incomplete that at best we can only reason on probabilities and possibilities.

CHAPTER EIGHT

OBSERVATION

"Knowledge comes from noticing resemblances and recurrences in the events that happen around us."
—WILFRED TROTTER

Illustrations

PASTEUR was curious to know how anthrax persists endemically, recurring in the same fields, sometimes at intervals of several years. He was able to isolate the organisms from soil around the graves in which sheep dead of the disease had been buried as long as 12 years before. He was puzzled as to how the organism could resist sunlight and other adverse influences so long. One day while walking in the fields he noticed a patch of soil of different colour from the rest and asked the farmer the reason. He was told that sheep dead of anthrax had been buried there the previous year.

"Pasteur, who always examined things closely, noticed on the surface of the soil a large number of worm castings. The idea then came to him that in their repeated travelling from the depth to the surface, the worms carried to the surface the earth rich in humus around the carcase, and with it the anthrax spores it contained. Pasteur never stopped at ideas but passed straight to the experiment. This justified his forecast. Earth contained in a worm, inoculated into a guinea-pig produced anthrax." [78]

This is a fine example of the value of direct personal observation. Had Pasteur done his thinking in an armchair it is unlikely that he would have cleared up this interesting bit of epidemiology.

When some rabbits from the market were brought into Claude Bernard's laboratory one day, he noticed that the urine which they passed on the table was clear and acid instead of turbid and alkaline as is usual with herbivorous animals. Bernard reasoned that perhaps they were in the nutritional condition of carnivora from having fasted and drawn on their own tissues for susten-

96

ance. This he confirmed by alternately feeding and starving them, a process which he found altered the reaction of their urine as he had anticipated. This was a nice observation and would have satisfied most investigators, but not Bernard. He required a " counterproof ", and so fed rabbits on meat. This resulted in an acid urine as expected, and to complete the experiment he carried out an autopsy on the rabbits. To use his words :

> " I happened to notice that the white and milky lymphatics were first visible in the small intestine at the lower part of the duodenum, about 30 cm. below the pylorus. The fact caught my attention because in dogs they are first visible much higher in the duodenum just below the pylorus."

On observing more closely, he saw that the opening of the pancreatic duct coincided with the position where the lymphatics began to contain chyle made white by emulsion of the fatty materials. This led to the discovery of the part played by pancreatic juice in the digestion of fats.[15]

Darwin relates an incident illustrating how he and a colleague failed to observe certain unexpected phenomena when they were exploring a valley :

> " Neither of us saw a trace of the wonderful glacial phenomena all around us; we did not notice plainly scored rocks, the perched boulders, the lateral and terminal moraines."[28]

These things were not observed because they were not expected or specifically looked for.

While watching the movements of the bacteria which cause butyric acid fermentation, Louis Pasteur noticed that when the organisms came near the edge of the drop they stopped moving. He guessed this was due to the presence of oxygen in the fluid near the air. Puzzling over the significance of this observation he concluded that there was no free oxygen where the bacteria were actively moving. From this he made the far reaching deduction that life can exist without oxygen, which at that time was thought not possible. Further he postulated that fermentation is a metabolic process by which microbes obtain oxygen from organic substances. These important ideas which Pasteur later substantiated had their origin in the observation of a detail that many would not have noticed.

Many of the anecdotes cited in Chapters Three and Four and in the Appendix also provide illustrations of the rôle of observation in research.

Some general principles in observation

In discussing the thoroughly unreliable nature of eye-witness observation of everyday events, W. H. George says:

> " What is observed depends on who is looking. To get some agreement between observers they must be paying attention, their lives must not be consciously in danger, their prime necessities of life must preferably be satisfied and they must not be taken by surprise. If they are observing a transient phenomenon, it must be repeated many times and preferably they must not only look *at*, but must look *for*, each detail."[47]

As an illustration of the difficulty of making careful observations, he tells the following story.

At a congress on psychology at Göttingen, during one of the meetings, a man suddenly rushed into the room chased by another with a revolver. After a scuffle in the middle of the room a shot was fired and both men rushed out again about twenty seconds after having entered. Immediately the chairman asked those present to write down an account of what they had seen. Although the observers did not know it at the time, the incident had been previously arranged, rehearsed and photographed. Of the forty reports presented, only one had less than 20 per cent mistakes about the principal facts, 14 had from 20 to 40 per cent mistakes, and 25 had more than 40 per cent mistakes. The most noteworthy feature was that in over half the accounts, 10 per cent or more of the details were pure inventions. This poor record was obtained in spite of favourable circumstances, for the whole incident was short and sufficiently striking to arrest attention, the details were immediately written down by people accustomed to scientific observation and no one was himself involved. Experiments of this nature are commonly conducted by psychologists and nearly always produce results of a similar type.

Perhaps the first thing to realise about observations is that not only do observers frequently miss seemingly obvious things, but what is even more important, they often invent quite false observations. False observations may be due to illusions, where the senses give wrong information to the mind, or the errors may have their origin in the mind.

Illustrations of optical illusions can be provided from various geometrical figures (see, for example, George[47]) and by distortions caused by the refraction of light when it passes through water, glass or heated air. Remarkable demonstrations of the unreliability of visual observations are provided by the tricks of " magicians " and conjurors. Another illustration of false information arising from the sense organs is provided by placing one hand in hot water and one in cold for a few moments and then plunging them both into tepid water. A curious fallacy of this nature was recorded by the ancient Greek historian, Herodotus :

> " The water of this stream is lukewarm at early dawn. At the time when the market fills it is much cooler; by noon it has grown quite cold; at this time therefore they water their gardens. As the afternoon advances, the coldness goes off, till, about sunset the water is once more lukewarm."

In all probability the temperature of the water remained constant and the change noticed was due to the *difference* between water and atmospheric temperatures as the latter changed. Fallacious observations of a similar type can be shown to arise from illusions associated with sound.

The second class of error in registering and reporting observation has its origin in the mind itself. Many of these errors can be attributed to the fact that the mind has a trick of unconsciously filling in gaps according to past experience, knowledge and conscious expectations. Goethe has said :

> " We see only what we know."

" We are prone to see what lies behind our eyes rather than what appears before them," an old saying goes. An illustration of this is seen in the cinema film depicting a lion chasing a negro. The camera shows now the lion pursuing, now the man fleeing, and after several repetitions of this we finally see the lion leap on something in the long grass. Even though the lion and the man may have at no time appeared on the screen together, most people in the audience are convinced they actually saw the lion

99

leap on the man, and there have been serious protests that natives were sacrificed to make such a film. Another illustration of the subjective error is provided by the following anecdote. A Manchester physician, while teaching a ward class of students, took a sample of diabetic urine and dipped a finger in it to taste it. He then asked all the students to repeat his action. This they reluctantly did, making grimaces, but agreeing that it tasted sweet. " I did this," said the physician with a smile, " to teach you the importance of observing detail. If you had watched me carefully you would have noticed that I put my first finger in the urine but licked my second finger !"

It is common knowledge that different people viewing the same scene will notice different things according to where their interests lie. In a country scene a botanist will notice the different species of plants, a zoologist the animals, a geologist the geological structures, a farmer the crops, farm animals, etc. A city dweller with none of these interests may see only a pleasant scene. Most men can pass a day in the company of a woman and afterwards have only the vaguest ideas about what clothes she wore, but most women after meeting another woman for only a few minutes could describe every article the other was wearing.

It is quite possible to see something repeatedly without registering it mentally. For example, a stranger on arrival in London commented to a Londoner on the eyes that are painted on the front of many buses. The Londoner was surprised, as he had never noticed them. But after his attention had been called to them, during the next few weeks he was conscious of these eyes nearly every time he saw a bus.

Changes in a familiar scene are often noticed even though the observer may not have been consciously aware of the details of the scene previously. Indeed sometimes an observer may be aware that something has changed in a familiar scene without being able to tell what the change is. Discussing this point, W. H. George says :

" It seems as if the memory preserves something like a photographic negative of a very familiar scene. At the next examination this memory image is unconsciously placed over the visual image present, and, just as with two similar photographic nega-

SIR ALEXANDER FLEMING

SIR HOWARD FLOREY

G. S. WILSON

F. M. BURNET

MAX PLANCK

SIR RONALD FISHER

C. H. ANDREWES

J. B. CONANT

tives, attention is immediately attracted to the places where the two do not exactly fit, that is, where there is a change in one relative to the other. It is noteworthy that this remembered whole cannot always be recalled to memory so as to enable details to be described."[47]

This analogy should not be taken too literally because the same phenomenon is seen with memory of other things such as stories or music. A child who is familiar with a story will often call attention to slight variations when it is retold even though he does not know it by heart himself. George continues:

" The perception of change seems to be a property of all of the sense organs, for changes of sound, taste, smell and temperature are readily noticed. . . . It might almost be said that a continuous sound is only 'heard' when it stops or the sound changes."[47]

If we consider that the comparison of the old and new images takes place in the subconscious, we can draw an analogy with the hypothesis as to how intuitions gain access to the conscious mind. One would expect the person to become aware of the notable facts, that is, the changes, even though he may be unable to bring all the details into consciousness.

It is important to realise that observation is much more than merely seeing something; it also involves a mental process. In all observations there are two elements: (a) the sense-perceptual element (usually visual) and (b) the mental, which, as we have seen, may be partly conscious and partly unconscious. Where the sense-perceptual element is relatively unimportant, it is often difficult to distinguish between an observation and an ordinary intuition. For example, this sort of thing is usually referred to as an observation: " I have noticed that I get hay fever whenever I go near horses." The hay fever and the horses are perfectly obvious, it is the connection between the two that may require astuteness to notice at first, and this is a mental process not distinguishable from an intuition. Sometimes it is possible to draw a line between the noticing and the intuition, e.g. Aristotle commented that on *observing* that the bright side of the moon is always toward the sun, it may suddenly *occur* to the observer that the explanation is that the moon shines by the light of the sun.

Similarly in three of the anecdotes given at the beginning of this chapter, the observation was followed by an intuition.

Scientific observation

We have seen how unreliable an observer's report of a complex situation often is. Indeed, it is very difficult to observe and describe accurately even simple phenomena. Scientific experiments isolate certain events which are observed by the aid of appropriate techniques and instruments which have been developed because they are relatively free from error and have been found to give reproducible results which are in accord with the general body of scientific knowledge. Claude Bernard distinguished two types of observation : (a) spontaneous or passive observations which are unexpected; and (b) induced or active observations which are deliberately sought, usually on account of an hypothesis. It is the former in which we are chiefly interested here.

Effective spontaneous observation involves firstly noticing some object or event. The thing noticed will only become significant if the mind of the observer either consciously or unconsciously relates it to some relevant knowledge or past experience, or if in pondering on it subsequently he arrives at some hypothesis. In the last section attention was called to the fact that the mind is particularly sensitive to changes or differences. This is of use in scientific observation, but what is more important and more difficult is to observe (in this instance mainly a mental process) resemblances or correlations between things that on the surface appeared quite unrelated. The quotation from Trotter at the beginning of this chapter refers to this point. It required the genius of Benjamin Franklin to see the relationship between frictional electricity and lightning. Recently veterinarians have recognised a disease of dogs which is manifest by encephalitis and hardening of the foot pads. Many cases of the disease have probably been seen in the past without anyone having noticed the surprising association of the encephalitis with the hard pads.

One cannot observe everything closely, therefore one must discriminate and try to select the significant. When practising a branch of science, the " trained " observer deliberately looks for specific things which his training has taught him are significant, but in research he often has to rely on his own discrimination, guided only by his general scientific knowledge, judgment and perhaps an hypothesis which he entertains. As Alan Gregg, the Director of Medical Sciences for the Rockefeller Foundation has said :

> "Most of the knowledge and much of the genius of the research worker lie behind his selection of what is worth observing. It is a crucial choice, often determining the success or failure of months of work, often differentiating the brilliant discoverer from the . . . plodder."[48]

When Faraday was asked to watch an experiment, it is said that he would always ask what it was he had to look for but that he was still able to watch for other things. He was following the principle enunciated in the quotation from George in the preceding section, that preferably each detail should be looked for. However, this is of little help in making original observations. Claude Bernard considered that one should observe an experiment with an open mind for fear that if we look only for one feature expected in view of a preconceived idea, we will miss other things. This, he said, is one of the greatest stumbling blocks of the experimental method, because, by failing to note what has not been foreseen, a misleading observation may be made. "Put off your imagination," he said, "as you take off your overcoat when you enter the laboratory." Writing of Charles Darwin, his son tells us that :

> "He wished to learn as much as possible from an experiment so he did not confine himself to observing the single point to which the experiment was directed. and his power of seeing a number of things was wonderful. . . . There was one quality of mind which seemed to be of special and extreme advantage in leading him to make discoveries. It was the power of never letting exceptions pass unnoticed."[23]

If, when we are experimenting, we confine our attention to only those things we expect to see, we shall probably miss the unexpected occurrences and these, even though they may at first be disturbing and troublesome, are the most likely to point the way to important unsuspected facts. It has been said that it is the exceptional phenomenon which is likely to lead to the explanation of the usual. When an irregularity is noticed, look for something with which it might be associated. In order to make original observations the best attitude is not to concentrate exclusively on the main point but to try and keep a look-out for the unexpected, remembering that observation is not passively watching but is an active mental process.

Scientific observation of objects calls for the closest possible scrutiny, if necessary with the aid of a lens. The making of detailed notes and drawings is a valuable means of prompting one to observe accurately. This is the main reason for making students do drawings in practical classes. Sir MacFarlane Burnet has autopsied tens of thousands of mice in the course of his researches on influenza, but he examines the lungs of every mouse with a lens and makes a careful drawing of the lesions. In recording scientific observations one should always be as precise as possible.

Powers of observation can be developed by cultivating the habit of watching things with an active, enquiring mind. It is no exaggeration to say that well developed habits of observation are more important in research than large accumulations of academic learning. The faculty of observation soon atrophies in modern civilisation, whereas with the savage hunter it may be strongly developed. The scientist needs consciously to develop it, and practical work in the laboratory and the clinic should assist in this direction. For example, when observing an animal, one should look over it systematically and consciously note, for instance, breed, age, sex, colour markings, points of conformation, eyes, natural orifices, whether the abdomen is full or empty, the mammary glands, condition of the coat, its demeanour and movements, any peculiarities and note its surroundings including any fæces or traces of food. This is, of course, apart from, or preliminary to, a clinical examination if the animal is ill.

In carrying out any observation you look deliberately for each characteristic you know may be there, for any unusual feature, and especially for any suggestive associations or relationships among the things you see, or between them and what you know. By this last point I mean such things as noticing that on a plate culture some bacterial colonies inhibit or favour others in their vicinity, or in field observations any association between disease and type of pasture, weather or system of management. Most of the relationships observed are due to chance and have no significance, but occasionally one will lead to a fruitful idea. It is as well to forget statistics when doing this and consider the possibility of some significance behind slender associations in the observed data, even though they would be dismissed at a glance if regarded on a mathematical basis. More discoveries have arisen from intense observation of very limited material than from statistics applied to large groups. The value of the latter lies mainly in testing hypotheses arising from the former. While observing one should cultivate a speculative, contemplative attitude of mind and search for clues to be followed up.

Training in observation follows the same principles as training in any activity. At first one must do things consciously and laboriously, but with practice the activities gradually become automatic and unconscious and a habit is established. Effective scientific observation also requires a good background, for only by being familiar with the usual can we notice something as being unusual or unexplained.

SUMMARY

Accurate observation of complex situations is extremely difficult, and observers usually make many errors of which they are not conscious. Effective observation involves noticing something and giving it significance by relating it to something else noticed or already known; thus it contains both an element of sense-perception and a mental element.

It is impossible to observe everything, and so the observer has to give most of his attention to a selected field, but he should at the same time try to watch out for other things, especially anything odd.

CHAPTER NINE

DIFFICULTIES

" Error is all around us and creeps in at the least opportunity. Every method is imperfect."—CHARLES NICOLLE.

Mental resistance to new ideas

WHEN the great discoveries of science were made they
appeared in a very different light than they do now.
Previous ignorance on the subject was rarely recognised, for
either a blind eye was turned to the problem and people were
scarcely aware of its existence, or there were well accepted
notions on the subject, and these had to be ousted to make way
for the new conceptions. Professor H. Butterfield points out
that the most difficult mental act of all is to re-arrange a familiar
bundle of data, to look at it differently and escape from the
prevailing doctrine.[20] This was the great intellectual hurdle
that confronted such pioneers as Galileo, but in a minor form
it crops up with every important original discovery. Things
that are now quite easy for children to grasp, such as the
elementary facts of the planetary system, required the colossal
intellectual feat of a genius to conceive when his mind was
already conditioned with Aristotelian notions.

William Harvey's discovery of the circulation of the blood
might have been relatively easy but for the prevailing beliefs
that the blood ebbed and flowed in the vessels, that there were
two sorts of blood and that the blood was able to pass from
one side of the heart to the other. His first cause for dissatisfaction with these doctrines was his finding of the direction in
which the valves faced in the veins of the head and neck—a
small stubborn fact which the current hypothesis did not fit. He
dissected no fewer than eighty species of animals including rep-
tiles, crustaceans and insects, and spent many years on the investigation. The big difficulty in establishing the conception of the
circulation was the absence of any visible connection between

the terminal arteries and the veins, and he had to postulate the existence of the capillaries, which were not discovered until later. Harvey could not demonstrate the circulation, so had to leave it as an inference. He required courage to announce how much blood he calculated that the heart pumped out. Harvey himself wrote :

> " But what remains to be said about the quantity and source of the blood which thus passes, is of so novel and unheard-of character that I not only fear injury to myself from the envy of a few, but I tremble lest I have mankind at large for my enemies, so much doth want and custom, that become as another nature, and doctrine once sown and that hath struck deep root, and respect for antiquity, influence all men: still the die is cast, and my trust is in my love of truth, and the candour that inheres in cultivated minds." [105]

His fears were well founded for he was subjected to derision and abuse and his practice suffered badly. Only after a struggle of over twenty years did the circulation of the blood become generally accepted.

Other illustrations of resistance to new ideas are provided by the stories about Jenner and Mules already recounted and that about Semmelweis given later in this chapter.

Vesalius in his early anatomical studies related that he could hardly believe his own eyes when he found structures not in accord with Galen's descriptions. Lesser men did, in fact, disbelieve their own eyes, or at least thought that the subject for dissection or their own handiwork was at fault. It is often curiously difficult to recognise a new, unexpected fact, even when obvious. Only people who have never found themselves face to face with a new fact laugh at the inability of medieval observers to believe their own eyes. Teachers well know that students often ignore the results of their experiments and mistrust their observations if they do not coincide with their expectations.

In nearly all matters the human mind has a strong tendency to judge in the light of its own experience, knowledge and prejudices rather than on the evidence presented. Thus new ideas are judged in the light of prevailing beliefs. If the ideas are too revolutionary, that is to say, if they depart too far from reigning theories and cannot be fitted into the current body of knowledge, they will not be acceptable. When discoveries are made before their time they are almost certain to be ignored or meet with opposition which is too strong to be overcome, so in most instances they may as well not have been made. Dr. Marjory Stephenson likens discoveries made in advance of their time to long salients in warfare by which a position may be captured. If, however, the main army is too far behind to give necessary support, the advance post is lost and has to be re-taken at a later date.[89]

McMunn discovered cytochrome in 1886, but it meant little and was ignored until Keilin rediscovered it thirty-eight years later and was able to interpret it. Mendel's discovery of the basic principles of genetics is another good example of inability of even the scientific world always to recognise the importance of a discovery. His work established the foundation of a new science, yet it was ignored for thirty-five years after it had been read to a scientific society and published. Fisher has said that each generation seems to have found in Mendel's paper only what it expected to find and ignored what did not conform to its own expectations.[38] His contemporaries saw only a repetition of hybridisation experiments already published, the next generation appreciated the importance of his views on inheritance but considered them difficult to reconcile with evolution. And now Fisher tells us that some of Mendel's results when examined in the hard cold light of modern statistical methods show unmistakable evidence of being not entirely objective—of being biased in favour of the expected result!

The work of some psychologists on extrasensory perception and precognition may be a present-day example of a discovery before its time. Most scientists have difficulty in accepting the conclusions of these workers despite apparently irrefutable evidence, because the conclusions cannot be reconciled with present knowledge of the physical world.

Unless made by someone outside accepted scientific circles, discoveries made when the time is ripe for them are more readily accepted because they fit into and find support in prevailing concepts, or indeed, grow out of the present body of knowledge. This type of discovery is bound to occur as part of the main current of the evolution of science and may arise more or less simultaneously in different parts of the world. Tyndall said :

> " Before any great scientific principle receives distinct enunciation by individuals, it dwells more or less clearly in the general scientific mind. The intellectual plateau is already high, and our discoverers are those who, like peaks above the plateau, rise a little above the general level of thought at the time."[95]

Such discoveries, nevertheless, often encounter some resistance before they are generally accepted.

There is in all of us a psychological tendency to resist new ideas which come from without just as there is a psychological resistance to really radical innovations in behaviour or dress. It perhaps has its origin in that inborn impulse which used to be spoken of as the herd instinct. This so-called instinct drives man to conform within certain limits to conventional customs and to oppose any considerable deviation from prevailing behaviour or ideas by other members of the herd. Conversely, it gives widely held beliefs a spurious validity irrespective of whether or not they are founded on any real evidence. Instinctive behaviour is usually rationalised, but the " reasons " are only secondary, being formed by the mind to justify its opinions. Wilfred Trotter said :

> " The mind likes a strange idea as little as the body likes a strange protein and resists it with similar energy. It would not perhaps be too fanciful to say that a new idea is the most quickly acting antigen known to science. If we watch ourselves honestly we shall often find that we have begun to argue against a new idea even before it has been completely stated."[94]

When adults first become conscious of something new they usually either attack or try to escape from it.[47] This is called the " attack-escape " reaction. Attack includes such mild forms as ridicule, and escape includes merely putting out of mind. The attack on the first man to carry an umbrella in London was an exhibition of the same reaction as has so often been displayed toward startling new discoveries in science. These attacks are often accompanied by rationalisations—the attacker giving the " reasons " why he attacks or rejects the idea. Scepticism is often an automatic reaction to protect ourselves against a new idea. How often do we catch ourselves automatically resisting a new idea someone presents to us. As Walshe says, the itch to suffocate the infant idea burns in all of us.[101]

Dale describes the ridicule which greeted Röntgen's first announcement of his discovery of X-rays.[27] An interesting feature of the story is that the great physicist J. J. Thomson did not share in the general scepticism, but on the contrary expressed a conviction that the report would prove to be true. Similarly, when Becquerel's discovery that uranium salts emitted radiations was announced, Lord Rayleigh was prepared to believe it while others were not. Thomson and Rayleigh had minds that were not enslaved by current orthodox views.

Some discoveries have had to be made several times before they were accepted. Writing of the resistance to new ideas Schiller says :

" One curious result of this inertia, which deserves to rank among the fundamental ' laws ' of nature, is that when a discovery has finally won tardy recognition it is usually found to have been anticipated, often with cogent reasons and in great detail. Darwinism, for instance, may be traced back through the ages to Heraclitus and Anaximander."[80]

It is not uncommon for opponents of an innovation to base their judgment on an " all or nothing " attitude, i.e., since it does not provide a complete solution to the practical problem, it is no use. This unreasonable attitude sometimes prevents or delays the adoption of developments which are very useful in the absence of anything better. We all know some scientists who steadfastly refuse to be convinced by the evidence in support of a discovery which conflicts with their preconceived ideas. Perhaps the persistent sceptic serves a useful purpose in the community, but I admit that it is not one which I admire. It is said that even today there are some people who still insist that the world is flat !

Nevertheless, exasperating and even harmful as resistance to discovery often is, it fulfils a function in buffering the community from the too hasty acceptance of ideas until they have been well proved and tried. But for this innate conservatism, wild ideas and charlatanry would be more rife than they are. Nothing could be more damaging to science than the abandonment of the critical attitude and its replacement by too ready acceptance of hypotheses put forward on slender evidence. The inexperienced scientist often errs in being too willing to believe plausible ideas. Superficially one's reaction to a new claim appears to be an example of the general problem of conservatism versus progressiveness. These attitudes of mind may subconsciously influence a person toward taking one side or the other in a dispute but we should strive against both of them. what we must aim at is honest, objective judgment of the evidence, freeing our minds as much as possible from opinion not based on fact, and suspend judgment where the evidence is incomplete. There is a very important distinction between a critical attitude of mind (or critical " faculty ") and a sceptical attitude.

Opposition to discoveries

Hitherto we have been concerned with psychological resistance to new ideas. In this section we will discuss some other aspects of opposition to discoveries.

Innovations are often opposed because they are too disturbing to entrenched authority and vested interests in the widest sense of that term. Zinsser quotes Bacon as saying that the dignitaries who hold high honours for past accomplishments do not usually like to see the current of progress rush too rapidly out of their reach. Zinsser comments :

> " Our task, as we grow older in a rapidly advancing science, is to retain the capacity of joy in discoveries which correct older ideas, and to learn from our pupils as we teach them. That is the only sound prophylaxis against the dodo-disease of middle age." [108]

Trouble over innovations is sometimes aggravated by the personality of the discoverer. Discoverers are often men with little experience or skill in human relations, and less trouble would have arisen had they been more diplomatic. The fact that Harvey succeeded eventually in having his discovery recognised, and that Semmelweis failed, may be explained on this basis. Semmelweis showed no tact at all, but Harvey dedicated his book to King Charles, drawing the parallel between the King and realm, and the heart and body. His biographer,

Willis, says he possessed in a remarkable degree the power of persuading and conciliating those with whom he came in contact. Harvey said:

> "Man comes into the world naked and unarmed, as if nature had destined him for a social creature and ordained that he should live under equitable laws and in peace; as if she had desired that he should be guided by reason."

In discussing his critics he remarked:

> "To return evil speaking with evil speaking, however, I hold to be unworthy in a philosopher [i.e. scientist] and searcher after the truth."[105]

Writing on the same subject Michael Faraday said:

> "The real truth never fails ultimately to appear: and opposing parties, if wrong, are sooner convinced when replied to forbearingly than when overwhelmed."[95]

The discoverer requires courage, especially if he is young and inexperienced, to back his opinion about the significance of his finding against indifferences and scepticism of others and to pursue his investigations. We take joy in reading of the courage displayed by men like Harvey, Jenner, Semmelweis and Pasteur in the face of opposition, but how often have profitable lines of investigation been dropped and lost in oblivion when the discoverer lacked the necessary zeal and courage? Trotter relates the story of J. J. Waterston who in 1845 wrote a paper on the molecular theory of gases anticipating much of the work of Joule, Clausius and Clerk Maxwell. The referee of the Royal Society to whom the paper was submitted said: "The paper is nothing but nonsense", and the work lay in utter oblivion until exhumed forty-five years later. Waterston lived on disappointed and obscure for many years and then mysteriously disappeared leaving no sign. As Trotter remarks, this story must strike a chill upon anyone impatient for the advancement of knowledge. Many discoveries must have thus been stillborn or smothered at birth. We know only those that survived.

Although in most countries to-day there is no risk attached to pursuing what are now orthodox scientific fields, it would be wrong to conclude that obscurantism and reaction are things only of the past. Barely thirty years ago Einstein suffered a virulent and organised campaign of persecution and ridicule in Germany[45] and in U.S.A. in 1925, at the notorious "Tennessee monkey trial", a science teacher was prosecuted for teaching evolution. In totalitarian states, the intrusion of politics into scientific matters, as was seen under the Nazi regime and now in Russia over the genetics controversy, may introduce authoritarianism into science with consequent suppression of the work of those not willing to bow to the party dictum on scientific theories.[5] A mild form of reaction persists in societies devoted to combating vaccination and vivisection. Nor should we scientists ourselves be too complacent, for even within scientific circles to-day a new discovery may be ignored or opposed if it is revolutionary in principle and made by someone outside approved circles. The discoverer may still be required to show the courage of his convictions.

It has been said that the reception of an original contribution to knowledge may be divided into three phases : during the first it is ridiculed as not true, impossible or useless; during the second, people say there may be something in it but it would never be of any practical use; and in the third and final phase, when the discovery has received general recognition, there are usually people who say that it is not original and has been anticipated by others.* Theobald Smith spoke truly when he said :

" The joy of research must be found in doing, since every other harvest is uncertain."[86]

It is a commonplace that in the past the great scientists have often been rewarded for their gifts to mankind by persecution. A good example of this curious fact is provided by the following story of what happened to Ignaz Semmelweis, when he showed how the dreadful suffering and loss of life due to puerperal fever that was then the rule in the hospitals of Europe could be prevented.

In 1847 Semmelweis got the idea that the disease was carried to the women on the hands of the medical teachers and students coming direct from the post-mortem room. To destroy the " cadaveric material " on the hands he instituted a strict routine

* This saying seems to have originated from Sir James Mackenzie (*The Beloved Physician*, by R. M. WILSON, John Murray, London).

of washing the hands in a solution of chlorinated lime before the examination of the patients. As a result of this procedure, the mortality from puerperal fever in the first obstetric clinic of the General Hospital of Vienna fell immediately from 12 per cent to 3 per cent, and later almost to 1 per cent. His doctrine was well received in some quarters and taken up in some hospitals, but such revolutionary ideas, incriminating the obstetricians as the carriers of death, roused opposition from entrenched authority and the renewal of his position as assistant was refused. He left Vienna and went to Budapest where he again introduced his methods with success. But his doctrine made little headway and was even opposed by so great a man as Virchow. He wrote a book, the famous *Etiology*, to-day recognised as one of the classics of medical literature; but then he could not sell it. Frustration made him bitter and irascible and he wrote desperate articles denouncing as murderers those who refused to adopt his methods. These met only with ridicule and finally he came to a tragic end in a lunatic asylum in 1865. Mercifully and ironically a few days after entering the asylum he died from an infected wound received in the finger during his last gynaecological operation : a victim of the infection to the prevention of which his whole life had been devoted. His faith that the truth of his doctrine would ultimately prevail was never shaken. In a rather pathetic foreword to his *Etiology* he wrote :

> " When I look back upon the past, I can only dispel the sadness which falls upon me by gazing into that happy future when the infection will be banished. But if it is not vouchsafed to me to look upon that happy time with my own eyes . . . the conviction that such a time must inevitably sooner or later arrive will cheer my dying hour."

The work of others, especially Tarnier and Pasteur in France and Lister in England, forced the world reluctantly to recognise, some ten years or more later, that what Semmelweis had taught was correct.

Semmelweis' failure to convince most people was probably because there was no satisfactory explanation of the value of disinfecting hands until bacteria were shown to be the cause of disease, and probably also because he did not exercise any diplomacy or tact. It is not clear that Semmelweis' efforts had much, or indeed any, influence on the final acceptance of the principles he discovered. Others seem to have solved the problem independently.[84]

Errors of interpretation

For want of a more appropriate place, I shall mention here some of the commoner pitfalls which are encountered in interpreting observations or experimental results and which have not already been discussed.

The most notorious source of fallacy is probably *post hoc, ergo propter hoc,* that is, to attribute a causal relationship between what has been done and what follows, especially to conclude in the absence of controls that the outcome has been influenced by some interference. All our actions and reason are based on the legitimate assumption that all events have their cause in what has gone before, but error often arises when we attribute a causal rôle to a particular preceding event or interference on our part which in reality had no influence on the outcome observed. The faith which the lay public has in medicines is due in a large measure to this fallacy. Until very recently the majority of medicines were of negligible value and had little or no influence on the course of the illness for which they were taken, nevertheless, many people firmly believed when they recovered that the medicine had cured them. A lot of people including some doctors, are convinced that certain bacterial vaccines prevent the common cold, because by a fortunate coincidence some patients had no cold the year following vaccination. Yet all the many controlled experiments done with similar vaccines failed to show the least benefit. The controlled experiment is the only way of avoiding this type of fallacy.

Much the same logical fallacy is involved in wrongly assuming that when an association between two events is demonstrated, the relationship is necessarily one of cause and effect. Sometimes data are collected which show that the incidence of a certain disease in a quarter of a city which is very smoky, or which is very low-lying, is much higher than in other quarters. The author may conclude that the smoke or low-lying ground predisposes to the disease. Often such conclusions are quite unjustified, and the cause should probably be sought in the poverty and overcrowding which is to be found in these insalubrious areas. Virchow, in refuting Semmelweis' doctrine about the causation of puerperal fever, asserted that the weather played an important part, because the highest incidence occurred in winter. Semmelweis replied that the association between epidemics and winter was due to the fact that it was in winter that the midwifery students spent most time on the dissection of dead bodies.

False conclusions can be drawn by attributing a causal rôle to a newly introduced factor whereas, in fact, the cause lies in the withdrawal of the factor which was replaced. Tests carried out among people accustomed to drinking coffee at night could show that a better night's sleep was obtained when a proprietary drink was taken instead of coffee. It might be claimed that the proprietary drink induced sleep whereas the better sleep might well be entirely due to coffee not having been taken. Similarly, false conclusions in dietetic experiments have sometimes been drawn when a new constituent has replaced another. The supposed effect of the new constituent has later proved to be due to the absence of the article of diet displaced. It was found that the blooming of some plants was influenced by supplementing day light with artificial light. At first this was thought to be due to the prolonged "day", but subsequently it was found to be due to the shortened "night", for breaking into the night with a brief period of illumination at midnight, was even more effective than a longer period of illumination near the evening or morning.

There is always a risk in applying conclusions reached from experimentation in one species, to another species. Many mistakes were made in concluding that man or a domestic animal required this or that vitamin because rats or other experimental animals did, but nowadays the error of this is generally appreciated. More recently the same trouble arose in chemotherapy. The sulphonamides which gave the best results in man were not always found to be the best against the same bacteria in some of the domestic animals.

A rather more insidious source of fallacy is failure to realise that there may be several alternative causes of one process.

W. B. Cannon[22] comments on the false deduction once made that adrenaline does not play a part in controlling the sugar level in the blood by calling forth sugar from the liver, on the ground that the blood-sugar level is maintained after removal of the adrenal medulla. The fact is that there are other methods of mobilising sugar reserves from the liver but none are so effective as adrenaline. Shivering by itself can prevent body temperatures from falling, but that does not prove that other processes cannot play a part. A variant of this "fallacy of a single cause" has been described by Winslow.[107] When a combination of two factors causes something, and one is universally present, it is usually rashly concluded that the other is the sole causal factor. In the nineteenth century it was believed that insanitary conditions in themselves caused enteric fever. The causal microbes were then universally present and the incidence of the disease was determined by presence or absence of sanitation. The cause of a disease is complex, consisting of a combination of causal microbe, the conditions necessary for its conveyance from one host to the next and factors affecting the susceptibility of the host. Any happening is the result of a complex of causal factors, one of which we usually single out as *the* cause owing to its not being commonly present as are the other circumstances.

Wrong conclusions about the incidence of some condition in a population are sometimes drawn through basing the observations on a section of the population which is not representative of the whole. For example, certain figures were generally accepted and printed in text-books as an index of the proportion of children at different ages that gave a negative reaction to the Schick test for immunity to diphtheria. Many years later these figures were found to be true only for children of the poorer classes attending public hospitals in the city. The figures for other sections of the population were very different. When I went to the U.S.A. in 1938, scarcely anyone I met could say a good word for President Roosevelt, but Dr. Gallup's method of sampling public opinion showed that more than fifty per cent supported him. There is a great temptation to generalise on one's own observations or experience, although often it is not based on a sample that is truly random or sufficiently large to be representative. Bacon warned against being led into error by relying on impressions.

> " The human understanding is most excited by that which strikes and enters the mind at once and suddenly, and by which the imagination is immediately filled and inflated. It then begins almost imperceptibly to conceive and suppose that everything is similar to the few objects which have taken impression on the mind."

A very common way in which mistakes arise is by making unjustified assumptions on incomplete evidence. To cite a classic example, in the lecture in which he enunciated his famous postulates, Robert Koch described how he had been led into error by making what appeared to be a reasonable assumption. In his pioneer work on the tubercle bacillus he obtained strains from a large variety of animal species and after having subjected them to a series of tests he concluded that all tubercle bacilli are similar. Only in the case of the fowl did he omit to do pathogenicity and cultural examinations because he could not at the time obtain fresh material. However, since the morphology was the same, he assumed that the organism from the fowl was the same as those from the other animals. Later he was sent several atypical strains of the tubercle bacillus which, despite a protracted investigation, remained a complete puzzle. He said :

> " When every attempt to discover the explanation of the discrepancy had failed, at length an accident cleared up the question."

He happened to get some fowls with tuberculosis and when he cultured the organisms from these :

> " I saw to my astonishment that they had the appearance and all the other characters of the mysterious cultures."

Thus it was he found that avian and mammalian tubercle bacteria are different.[57] Incidentally, this reference, which I found when looking for something else, seems to have been " lost ", for some current text-books state that there is no evidence that Koch ever put forward the well-known postulates contained in this lecture.

One can easily be led astray when attempting to isolate an infective agent by inoculation and passage in experimental animals. Many mice carry in their nose latent viruses which, when any material is inoculated into the lungs through the nose, are carried into the lungs where they multiply. If the lungs from these mice are used to inoculate other mice in the same way, pneumonia is sometimes set up and, as a result, it might be wrongly concluded that a virus had been isolated from the original material. Also in attempting to isolate a virus by inoculating material on to the skin of experimental animals, it is possible to set up a transmissible condition which originated from the environment and not from the original inoculum.

Early investigations on distemper of dogs incriminated as the causal agent a certain bacterium isolated from cases of the disease because on inoculation it set up a disease resembling distemper. When later a virus was shown to be the true cause of the distemper, it became apparent that the early investigators had been misled either because they had isolated a pathogenic secondary invader or because they had not taken sufficiently rigid measures to quarantine their experimental dogs.

When the investigator has done his best to detect any errors in his work, a service that colleagues are usually glad to assist with is criticism. He is a bold man who submits his paper for publication without it having first been put under the microscope of friendly criticism by colleagues.

SUMMARY

The mental resistance to new ideas is partly due to the fact that they have to displace established ideas. New facts are not usually accepted unless they can be correlated with the existing body of knowledge; it is often not sufficient that they can be demonstrated on independent evidence. Therefore premature discoveries are usually neglected and lost. An unreasoning, instinctive mental resistance to novelty is the real basis of excessive scepticism and conservatism.

Persecution of great discoverers was due partly to mental resistance to new ideas and partly to the disturbance caused to entrenched authority and vested interests, intellectual and material. Sometimes lack of diplomacy on the part of the discoverer has aggravated matters. Opposition must have killed at birth many discoveries. Obscurantism and authoritarianism are not yet dead.

Included among the many possible sources of fallacy are *post hoc, ergo propter hoc*, comparing groups separated by time, assuming that when two factors are correlated the relationship is necessarily one of cause and effect, and generalising from observations on samples that are not representative.

CHAPTER TEN

STRATEGY

" Work, Finish, Publish."—Michael Faraday.

Planning and organising research

MUCH controversy has taken place over planning in research. The main disagreement is on the relative merits of pure and applied research, on what proportion of the research in a country should be planned and to what degree it should be planned. The extreme advocates of planning consider that the only research worth while is that which is undertaken in a deliberate attempt to meet some need of society, and that pure research is seldom more than an elegant and time-wasting amusement. On the other hand the anti-planners (in England there is a Society for Freedom in Science) maintain that the research worker who is organised becomes only a routine investigator because, with the loss of intellectual freedom, originality cannot flourish.

Discussions on planning research are often confused by failure to make clear what is meant by planning. It is useful to distinguish three different levels of planning. The first is the actual conduct of an investigation by the worker engaged in the problem. This corresponds with tactics in warfare. It is short term and seldom goes far beyond the next experiment. The second level involves planning further ahead on broad lines and corresponds with strategy in warfare. Planning at this level is not confined to the man engaged in the problem but is also often the concern of the research director and the technical committee. Finally there is planning of policy. This type of planning is mostly done by a committee which decides what problems should be investigated and what projects or workers should receive support.

It has already been pointed out that many discoveries are

quite unforeseen, and that the principal elements in biological research are intensely individual efforts in (*a*) recognising the unexpected discovery and following it up, and (*b*) concentrated prolonged mental effort resulting in the birth of ideas. Major discoveries probably result less frequently from the systematic accumulation of data along planned lines. It is not a fact, as some suppose, that no solution to a problem is likely to be found until we have fundamental knowledge on the subject. Frequently an empirical discovery is made providing a solution and the rationale is worked out afterwards. One of the principal morals to be drawn from the discoveries described in this book is that the research worker ought not, having decided on a course of action, to put on mental blinkers and, like a cart-horse, confine his attention to the road ahead and see nothing by the way.

In view of these lessons which are to be learnt from the history of scientific discovery, research is less likely to be fruitful where the investigation is planned at the tactical level by a committee than when the person actually doing the research works out his own tactics as the investigation unfolds. Research is for most workers an individualistic thing and the responsibility for tactical planning is best left to the individual workers, who will devote their mental energies to the subject if they are allowed the incentives and rewards that are essential for fruitful research. Initiative can be easily discouraged by too much supervision for a man will seldom put his whole heart into a problem unless he feels that it is his own. Simon Flexner, the founder of the Rockefeller Institute of Medical Research, always believed that men of the right sort could be trusted to have better ideas than others could think up for them.[77] The scientist should not even be expected to adhere in detail to a programme of work which he himself has drawn up, but should be allowed to vary it as developments require.

The late Professor W. W. C. Topley said :

"Committees are dangerous things that need most careful watching. I believe that a research committee can do one useful thing and one only. It can find the workers best fitted to attack a particular problem, bring them together, give them the facilities they need, and leave them to get on with the work. It can review progress from time to time, and make adjustments; but if it tries to do more, it will do harm."[92]

Technical committees and research directors can often help in planning at the strategic level providing they work in consultation with the man who is going to do the work and do not attempt to dictate tactics. Committees are of most value in planning at the policy level, in calling attention to problems of importance to the community and making available the necessary finances and scientists. Another useful function that a committee can sometimes perform is to accelerate advances by seeing that workers in different laboratories are kept informed of each other's progress without the usual delay entailed in publication. Some war-time committees did useful service in co-ordinating scattered work in this way.

It is perhaps so obvious as to be scarcely worth mentioning that planning at the strategic and policy levels places a heavy responsibility on the planners, and is only likely to be successful when entrusted to people who have a real understanding of research as well as a good general knowledge in science. It is generally recognised that a committee which draws up programmes of research at the strategic level should consist mainly of men actively engaged in the field of research in which the problem falls. Unfortunately often committees are too inclined to play safe and support only projects which are planned in detail and follow conventional lines of work. Worthwhile advances are seldom made without taking risks.

Plans and projects are in order for tackling recognised problems, that is to say, for applied research, but science also needs the independent worker who pursues pure research without thought of practical results.

In team work some individual or individuals should usually take the lead and do the thinking. There are, of course, some scientists who are not well fitted to do independent research and yet who may be very useful working under close direction as members of a team. Other things being equal, the person with a fertile imagination makes a better leader than someone with a purely logical mind, for the former is more inspiring as well as more useful in providing ideas. But the leader of a team needs to be actively engaged on the problem himself.

In other words the tactical planning is best done by the bench worker, not the office administrator. Where there is not an acknowledged leader of the team, the problem can often be divided up so that each person capable of independent work has his own aspect of the problem for which he is responsible. The thing to avoid is too detailed and rigid planning by the assembled team. However, when team work is undertaken, the work ought to be sufficiently co-ordinated for each to understand not only his own special aspect but have a good grasp of the problem as a whole. The principles of team work were well expressed by Ehrlich: "Centralisation of investigation with independence of the individual worker." All plans must be regarded as tentative and subject to revision as the work progresses. One must not confuse the planning of research with the planning of individual experiments. No one would dispute the advisability of devoting great care to the planning of experiments and carrying them through according to plan.

Team work is essential in research in the investigation of problems which overlap into several branches of science, for instance, the investigation of a disease by a clinician, bacteriologist and biochemist. Large teams are most frequently used in biochemical investigations where there is need for a large amount of co-ordinated skilled technical work. Also team work is often required to develop discoveries which have originated from individual workers.

Another important use of the team is to increase the capacity of the brilliant man beyond what he could do with only his hands and technical assistance. The research team, especially of this type, also is valuable in providing an opportunity for the beginner to learn to do research. The young scientist benefits more from working in collaboration with an experienced research worker than by only having supervision from him. Also in this way he is more likely to get a taste of success, which is a tremendous help. Moreover, the association of the freshness and originality of youth with the accumulated knowledge and experience of a mature scientist can be a mutually beneficial arrangement. Where close collaboration is involved, the personalities of the individuals are, of course, an important consideration. Most brilliant men are stimulating to others, but some are so full of ideas from their own fertile mind and are so keen to try them out that they have a cramping effect on a junior colleague who wants to try out his own ideas. Moreover, it is possible for a man to be a brilliant scientist and yet be quite undeveloped in the knowledge and practice of human personal relations.

The objection most often raised against team work is that those discoveries which arise from unexpected side issues will be missed if the worker is not free to digress from his investigation. Fleming has pointed out that had he been working in a team he would not have been able to drop what he was doing and follow the clue that led to penicillin.[42]

For his own guidance the research worker himself needs to make at least some tentative general plan of an investigation at the outset and to make very careful detailed plans for actual experiments. It is here that the experience of the research director can be most helpful to the young scientist. The latter presents for discussion a general picture of the information he has collected, together with his ideas for the proposed work. The inexperienced scientist usually does not realise the limitations of what is practicable in research, and often proposes for one year's work a plan that would occupy him for ten. The experienced man knows that it is a practical necessity to confine himself to a fairly simple project because he realises how much work even that entails. From hearing of only the successful investigations the uninitiated often gets a false idea of the easiness of research. Advances are nearly always slow and laborious and one person can attempt only a limited objective at a time. It is as well for the beginner to discuss with his supervisor any important deviations from the plan because although fruitful clues may arise which should be followed, it is neither possible nor desirable to pursue every unanswered question that comes up. To give advice on these issues and to help when difficulties are met are the main functions of a director of research, and the successes of those under his direction are a measure of his understanding of the nature of scientific investigation. As the young scientist develops he should be encouraged to become less and less dependent on his seniors. The rate at which this independence develops will be determined by the aptitude that he shows and the success he attains.

Both the team worker and the individual worker usually find it useful to keep a list of the ideas and experiments he intends to try—a work programme, which is revised continuously.

Some consider that the best work is done in small research institutes where the director can keep in intimate touch with all the work, and that when this size is passed efficiency drops. It is undoubtedly true that there are examples of small institutes whose output per man is better than in the average large institute. In such places one usually finds a director who is not only a capable scientist but who also stimulates enthusiasm in his staff. High productivity in large institutes perhaps depends on there being several active foci, each centred on a good leader.

Different types of research

Research is commonly divided into "applied" and "pure". This classification is arbitrary and loose, but what is usually meant is that applied research is a deliberate investigation of a problem of practical importance, in contradistinction to pure research done to gain knowledge for its own sake. The pure scientist may be said to accept as an act of faith that any scientific knowledge is worth pursuing for its own sake, and, if pressed, he usually claims that in most instances it is eventually found to be useful. Most of the greatest discoveries, such as the discovery of electricity, X-rays, radium and atomic energy, originated from pure research, which allows the worker to follow unexpected, interesting clues without the intention of achieving results of practical value. In applied research it is the project which is given support, whereas in pure research it is the man. However, often the distinction between pure and applied research is a superficial one as it may merely depend on whether or not the subject investigated is one of practical importance. For example, the investigation of the life cycle of a protozoon in a pond is pure research, but if the protozoon studied is a parasite of man or domestic animal the research would be termed applied. A more fundamental differentiation, which corresponds only very roughly with the applied and pure classification is (a) that in which the objective is given and the means of obtaining it are sought, and (b) that in which the discovery is first made and then a use for it is sought.

There exists in some circles a certain amount of intellectual snobbery and tendency to look disdainfully on applied investigation. This attitude is based on the following two false ideas: that new knowledge is only discovered by pure research while applied research merely seeks to apply knowledge already available, and that pure research is a higher intellectual activity because it requires greater scientific ability and is more difficult. Both these ideas are quite wrong. Important new knowledge has frequently arisen from applied investigation; for instance, the science of bacteriology originated largely from Pasteur's investigations of practical problems in the beer, wine and silkworm industries. Usually it is more difficult to get results in applied research than in pure research, because the worker has to stick to and solve a given problem instead of following any promising clue that may turn up. Also in applied research most fields have already been well worked over and many of the easy and obvious things have been done. Applied research should not be confused with the routine practice of some branch of science where only the application of existing knowledge is attempted. There is need for both pure and applied research for they tend to be complementary.

Practical problems very often require for their solution more than the mere application of existing knowledge. Frequently gaps in our knowledge are found that have to be filled in. Furthermore, if applied research is limited to finding a solution to the immediate problem without attempting to arrive at an understanding of the underlying principles, the results will probably be applicable only to the particular local problem and will not have a wide general application. This may mean that similar and related problems have to be investigated afresh, whereas had the original investigation been done properly it would have provided the solution to the others. Even an apparently simple matter such as the practical development of a discovery may present unsuspected difficulties. When the new insecticide, gammexane, was adopted for use as a sheep dipping fluid, very careful tests and field trials were conducted to determine that it was non-toxic and in every way harmless. But despite its having passed an extensive series of tests, when it became widely used in the field, sheep in a number of flocks developed severe lameness after dipping. Investigation showed that the lameness was not due to the gammexane but to infection with a certain bacterium. The dipping fluid had become fouled with this bacterium which was carried in by some of the sheep. Dipping fluids used previously had a germicidal action against this bacterium, but gammexane had not. Problems of control in biology are often different in different localities. The malaria parasite may have as an intermediate host a different species of mosquito and the liver fluke may utilise a different snail.

Applied research cuts horizontally across several pure sciences looking for newly found knowledge that can be used in the practical problem. However, the applied scientist is not content with waiting for the discoveries of the pure scientist, valuable as they are. The pure scientist leaves serious gaps in those aspects of the subject which do not appeal to him, and the applied scientist may have to initiate fundamental research in order to fill them.

Scientific research may also be divided into the exploratory type which opens up new territory, and developmental type which follows on the former. The exploratory type is free and adventurous; occasionally it gives us great and perhaps unexpected discoveries; or it may give us no results at all. Developmental type of research is more often carried on by the very methodical type of scientist who is content to consolidate the advances, to search over the newly won country for more modest discoveries, and to exploit fully the newly gained territory by putting it to use. This latter type of research is sometimes spoken of as "pot-boiling" or "safety first" research.

"Borderline" research is research carried on in a field where two branches of science meet. This can be very productive in the hands of a scientist with a sufficiently wide training because he can both use and connect up knowledge from each branch of science. A quite ordinary fact, principle or technique from one branch of science may be novel and fruitful when applied in the other branch.

Research may be divided into different levels which are reached successively as a branch of science or a subject becomes more advanced. First comes the observational type of research carried out by naturalists in the field or by scientists with similar mental attributes in the laboratory. Gradually the crude phenomena and materials become refined to more precise but more restricted laboratory procedures, and these ultimately are reduced to exact physical and chemical processes. It is almost a practical impossibility for anyone to have a specialist knowledge of more than a limited field at one level. The natural historian type, who is no less useful than his colleagues, owes most of his success to his powers of observation and natural wit and often lacks the depth of basic scientific knowledge necessary to develop his findings to the full. On the other hand, the specialist in a basic science may be too far removed, mentally and physically, from phenomena occurring in nature to be the equal of the natural historian type in starting new lines of work.

The transfer method in research

All scientific advances rest on a base of previous knowledge. The discoverers are the people who supply the keystone to another arch in the building and reveal to the world the completed structure built mainly by others. In this section, however, I am referring not so much to the background of knowledge on which one tries to build but rather to the adaptation of a piece of new knowledge to another set of circumstances.

Sometimes the central idea on which an investigation hinges is provided by the application or transfer of a new principle or technique which has been discovered in another field. The method of making advances in this way will be referred to as the "transfer" method in research. This is probably the most fruitful and the easiest method in research, and the one most employed in applied research. It is, however, not to be in any way despised. Scientific advances are so difficult to achieve that every useful stratagem must be used. Some of these contributions might be more correctly called developments rather than discoveries since no new principles and little new knowledge may be brought to light. However, usually in attempting to apply the newly discovered principle or technique to the different problem, some new knowledge does arise.

Transfer is one of the principal means by which science evolves. Most discoveries have applications in fields other than those in which they are made and when applied to these new fields they are often instrumental in bringing about further discoveries. Major scientific achievements have sometimes come from transfer. Lister's development of antiseptic surgery was largely a transfer of Pasteur's work showing that decomposition was due to bacteria.

It might be thought that as soon as a discovery is announced, all its possible applications in other fields follow almost immediately and automatically, but this is seldom so. Scientists sometimes fail to realise the significance which a new discovery in another field may have for their own work, or if they do realise it they may not succeed in discovering the necessary modifications. Years elapsed between the discovery of most of the principles of bacteriology and immunology and all their applications to various diseases. It was some time before the principle of haemagglutination by viruses, discovered by Hirst with influenza virus, was found to hold with several other viruses, however with modifications in some instances, as one might have expected, and still later it has been extended to certain bacteria.

An important form of the transfer method is the exploitation of a new technique adopted from another branch of science. Some workers deliberately take up a new technique and look for problems in which its special virtues offer new openings. Partition chromatography and haemagglutination have, for example, been used in this way in fields far removed from those in which they were first developed.

The possibility of developments by the transfer method is perhaps the main reason why the research man needs to keep himself informed of at least the principal developments taking place in more than his own narrow field of work.

In this section we might also mention the scientific developments of customs and practices already in use without any scientific background. A large number of drugs used in therapeutics came into use in this way. Quinine, cocaine, curare and ephedrine were used long before they were studied scientifically and their pharmacological action understood. The medicinal properties of the herb Ma Huang, from which ephedrine is derived are said to have been discovered in China, 5,000 years ago by the emperor Shen Nung. The discoveries of quinine, cocaine and curare by the natives in South America are lost in antiquity but obviously they must have been purely empirical. Incidentally, the tree from which quinine is obtained was named after the Countess of Cinchona who used it to cure malaria in 1638 and subsequently introduced it into Europe from Peru. Another example of this type of investigation is research into age-old processes such as tanning, cheese making and fermentation of various kinds. Many of these processes have now been developed into exact scientific procedures and thereby improved, or at least made more dependable. Vaccination could perhaps also be classified under this heading.

Tactics

In order to examine and get a better understanding of a complex process, it is often useful to analyse it into component phases and consider each separately. This is what has been done in this treatise on research. I have tried to describe the rôle of hypothesis, reason, experimentation, observation, chance and intuition in research and to indicate the special uses and defects of each of these factors. However, in practice these factors of course do not operate separately. Several or all are usually required in any investigation, although often the actual key to the solution of the problem is provided by one, as is shown in many of the anecdotes cited.

A general outline of how a straightforward problem in experimental medicine or biology may be tackled has been given in Chapters One and Two and the special rôle of each factor in research has been discussed in subsequent chapters. The order of the chapters has no special significance, nor does the space devoted to each subject bear much relationship to its relative importance. There remain to be discussed only some general considerations about tactics. In doing this it may be useful to recapitulate and bring together some of the points already made elsewhere.

No set rules can be followed in research. The investigator has to exercise his ingenuity, originality and judgment and take advantage of every useful stratagem. F. C. S. Schiller wrote:

> "Methods that succeed must have value. . . . The success has shown that *in this case* the enquirer was right to select the facts he fixed upon as significant, and to neglect the rest as irrelevant, to connect them as he did by the ' laws ' he applied to them, to theorise about them as he did, to perceive the analogies, to weigh the chances, as he did, to speculate and to run the risks he did. But only in this case. In the very next case, which he takes to be ' essentially the same ' as the last, and as nearly analogous as is humanly possible, he may find that the differences (which always exist between cases) are relevant, and that his methods and assumptions have to be modified to cope with it successfully."[80]

Research has been likened to warfare against the unknown. This suggests some useful analogies as to tactics. The first consideration is proper preparation by marshalling all available resources of data and information, as well as the necessary material and equipment. The attacker will have a great advantage if he can bring to bear a new technical weapon. The procedure most likely to lead to an advance is to concentrate one's forces on a very restricted sector chosen because the enemy is believed to be weakest there. Weak spots in the defence may be found by preliminary scouting or by tentative attacks; when a stiff resistance is encountered it is usually better to seek a way around it by some manoeuvre instead of persisting in a frontal attack. Very occasionally, when a really important break-through is effected, it may be expedient, although risky, to overrun quickly a large territory and leave much of the consolidation to followers, provided the work is important enough to attract them. However, generally speaking, advances proceed by stages; when a new position is taken it should be firmly consolidated before any attempt is made to use it as a base for further advance. This rhythm is the normal form of progression not only in scientific research but in all forms of scholarship: the gathering of information leads naturally to a pause for synthesis and interpretation which in turn is followed by another stage of collection of crude data selected in light of the new generalisations reached.

Even in applied research, such as the investigation of a disease of man or of domestic animals, the usual procedure is first to find out as much as possible about any or all of the aspects of the problem, without deliberately aiming at a particular objective of practical use. Experience has shown quite definitely that a fuller understanding of the subject nearly always reveals useful facts. Sometimes one finds a vulnerable link in the life-cycle of the parasite causing the disease and this may lead to a simple means of control. Having such a possibility in view it is helpful to consider the biology of the infective agent, whether it be virus or helminth, and to ponder on how it manages to survive, especially when making its way from one host to the next.

Biological discoveries are often at first recognised in the form of qualitative phenomena and one of the first aims is usually to refine them to quantitative, reproducible processes. Eventually they may be reduced to a chemical or physical basis. It is noteworthy that the declared aim of a large proportion of investigations described in the leading scientific journals is to disclose the mechanism of some biological process. It is a fundamental belief that all biological functions can eventually be explained in terms of physics and chemistry. Vitalism, which postulated mysterious " vital " forces, and teleology, which postulated a supernatural directing agency, have long ago been abandoned by experimental biologists. However, teleology is admissible in a modified sense that an organ or function fulfils a purpose toward aiding the survival of the organism as a whole or survival of the species.

The most honoured and acclaimed advances in science are the perception of new laws and principles and factual discoveries of direct practical use to man. Usually little prominence is given to the inventions of new laboratory techniques and apparatus despite the fact that the introduction of an important new technique is often responsible for a surge of progress just as much as is the discovery of a new law or fact. Solid media for the culture of bacteria, bacterial filters, virus haemagglutination and partition chromatography are outstanding examples. It may be profitable for research workers and the organisers of research to pay more attention to the developments of new techniques than has been the custom.

It was a characteristic of Faraday, Darwin, Bernard and

probably all great investigators to follow up their discoveries and not leave the trail till they had exhausted it. The story of Bernard's experiments with digestion in rabbits recounted earlier provides a good illustration of this policy. When Gowland Hopkins found that a certain test for proteins was due to the presence of glyoxylic acid as an impurity in one of the reagents, he followed this up to find what group in the protein it reacted with and this led to his famous isolation of tryptophane. Any new fact is potentially an important new tool to be used for uncovering further knowledge and a small discovery may lead to something much greater. As Tyndall said :

"Knowledge once gained casts a faint light beyond its own immediate boundaries. There is no discovery so limited as not to illuminate something beyond itself." [95]

As soon as anything new is discovered the successful scientist immediately looks at it from all possible points of view and by connecting it with other knowledge seeks new avenues for investigation. The real and lasting pleasure in a discovery comes not so much from the accomplishment itself as from the possibility of using it as a stepping stone for fresh advances.

Anyone with a spark of the research spirit does not need to be exhorted to chase for all he is worth a really promising clue when one is found, dropping for the time being other activities and interests as far as practicable. But in research most of the time progress is difficult and often one is up against what appears to be a "brick wall". It is here that all resources of ingenuity and method are required. Perhaps the first thing to try is to abandon the subject for a few days and then reconsider the whole problem with a fresh mind. There are three ways in which benefit may be derived from temporary abandonment of a difficulty. It allows time for "incubation", that is for the subconscious to digest the data, it allows time for the mind to forget conditioned thinking, and lastly, by not doggedly persisting, one avoids fixing too strongly the unprofitable lines of thought. The principle of temporary abandonment is, of course, widely practised in everyday life, as for example, in postponing the making of a difficult decision until one has "slept on it". Elsewhere the usefulness of discussion has been stressed, not so much for seeking technical advice as for promoting new ideas. Also discussion helps one to gain that clear understanding of the problem, which is so essential.

Another thing to try when one is up against an impasse is to go back to the beginning and try to find a new line of approach by looking at the problem in a different way. It may be possible to collect more data from the field or clinic. Fresh field or clinical observations may also be useful in prompting new ideas. As a result of trying to reduce the problem to an experimental inquiry, the worker may have selected a sterile and erroneous refinement of the problem. When the crude problem is seen again he may select some other aspect for investigation. Sometimes it is possible to resolve the difficulty into simpler components which can be tackled separately. If the difficulty cannot be overcome, perhaps a way around it can be found by using an alternative technical method. It may be helpful to look for analogies between the problem presented and others that have been solved.

If, after persistent attempts to resolve the difficulty, no advance is being made, it is usually best to drop the problem for a few weeks or months and take up something else, but to think and talk about it occasionally. A new idea may arise or a new development in other fields may occur which enable the problem to be taken up again. If nothing fresh turns up, the problem will have to be abandoned as being insoluble in the present state of knowledge in related fields. It is, however, a serious fault in a research worker to be too ready to drop problems as soon as he encounters a difficulty or gets seized by enthusiasm for another line of work. Generally speaking one should make every effort to complete an investigation once it has been started. The worker who repeatedly changes his problem to chase his newest bright idea is usually ineffectual.

As soon as a piece of work is nearing completion it should be written up as for publication. It is important to do this before the work has been brought to a close because frequently one finds gaps or weak points which can be remedied while the materials are still at hand. Even when the work is not nearing completion, it is as well to write up an investigation at least once a year, because otherwise when one writes up work from old notes, one's memory of the experiments has become dim so that the task is more difficult and cannot be done so well. Also, for reasons discussed elsewhere, it is desirable to review the problem periodically. However, work that has not produced significant results is better not published. It clutters up the journals and does more harm than good to the author's reputation in the minds of the discerning.

When the work has been completed, it is wise to submit the article to an experienced colleague for criticism—not only because the colleague may be more experienced than the author, but also because it is easier to see flaws in another's work or language than in one's own.

A word of caution might be given against publishing work that is not conclusive and especially about making interpretations that are not fully justified by the experimental results or observations. Whatever is written will remain permanently in the literature and one's scientific reputation can be damaged by publishing something that is later proved incorrect. Generally speaking, it is a safe policy to give a faithful record of the results obtained and to suggest only cautiously the interpretation, distinguishing clearly between facts and interpretation. Premature publication of work that could not be substantiated has at times spoilt the reputation of promising scientists. Superlatives and exaggeration are anathema to most scientists, the greatest of whom have usually been modest and cautious. Faraday wrote to a friend in 1831 :

" I am busy just now again on electro-magnetism, and think I have got hold of a good thing, but can't say. It may be a weed instead of a fish that, after all my labour, I may at last pull up."

What he pulled up was the electric dynamo. In 1940 Sir Howard Florey wrote to the Rockefeller Foundation for financial support for his work on penicillin, which he then had good reason for believing could be developed into a therapeutic agent even more effective than the sulphonamides. In such a letter one might be expected to present the work in the most favourable light, but this is all that Florey allowed himself to say :

" I don't think I am too optimistic in thinking that this is a very promising line."[76]

What a classic piece of understatement that has proved to be !

I confess that I did not read Bacon until after I had nearly finished writing this book and only then did I realise how clearly he had seen that discovery is more often than not empirical—the same view as I have reached from studying the methods which have produced results during recent times. He quotes with approval Celsus as saying:

> "That medicines and cures were first found out, and then after the reasons and causes were discoursed; and not the causes first found out, and by light from them the medicines and cures discovered."[6]

No more apt commentary could be made about the advances in chemotherapy of this century than this remark of Celsus' about the medical science of 1800 years ago. When one reflects that chance and empiricism is the method by which organic evolution developed, it is perhaps not so surprising that these factors play such an important part in biological research.

In research we often have to use our techniques at their extreme limit and even beyond—like Schaudinn discovering the pale spirochaete of syphilis which others could barely see by the methods then available. So also with our reasoning; for usually discovery is beyond the reach of reason.

In physics inductive logic is as inadequate as in biology. Einstein leaves us in no doubt on this point when he says:

> "There is no inductive method which could lead to the fundamental concepts of physics. Failure to understand this fact constituted the basic philosophical error of so many investigators of the nineteenth century. . . . We now realise with special clarity, how much in error are those theorists who believe that theory comes inductively from experience."

In formal education the student is implicitly, if not explicitly, led to believe that reason is the main, or even the only, means by which science advances. This view has been supported by the conception of the so-called "scientific method" outlined mainly by certain logicians of the last century who had little real understanding of research. In this book I have tried to show the error of this outlook and have emphasised the limitations of reason as an instrument in making discoveries. I have not questioned the belief that reason is the best guide in known territory, though even here the hazards in its use are probably greater than generally realised. But in research we are continually groping beyond known territory and here it is not so much a question of abandoning reason as finding that we are unable to employ it because there is not sufficient information available on which to use it properly. Rather than delude ourselves that we are able effectively to use reason in complex natural phenomena when we have only inadequate information and vague ideas, it seems to me better openly to recognise that we have often to resort to taste and to recognise the important rôles of chance and intuition in discovery.

In research, as indeed in everyday life, very often we have of necessity to decide our course of action on personal judgment based on taste. Only the technicalities of research are " scientific " in the sense of being purely objective and rational. Paradoxical as it may at first appear, the truth is that, as W. H. George has said, scientific research is an art, not a science.[47]

SUMMARY

Tactics are best worked out by the worker engaged on the problem. He should also have a say in planning strategy, but here he can often be assisted by a research director or by a technical committee which includes scientists familiar with the particular field of work. The main function of committees is planning matters of policy. Research can be planned but discovery cannot.

When discoveries are transferred to another field of science they are often instrumental in uncovering still further knowledge. I have given some hints on how best to go about the various activities that constitute research, but explicit rules cannot be laid down because research is an art.

The general strategy of research is to work with some clear object in view but nevertheless to keep alert for and seize any unexpected opportunities.

CHAPTER ELEVEN

SCIENTISTS

"It is not the talents we possess so much as the use we make of them that counts in the progress of the world."
BRAILSFORD ROBERTSON

Attributes required for research

IN MANY respects the research worker resembles the pioneer. He explores the frontiers of knowledge and requires many of the same attributes: enterprise and initiative, readiness to face difficulties and overcome them with his own resourcefulness and ingenuity, perseverance, a spirit of adventure, a certain dissatisfaction with well-known territory and prevailing ideas, and an eagerness to try his own judgment.

Probably the two most essential attributes for the research worker are a love of science and an insatiable curiosity. The person attracted to research usually is one who retains more than usual of the instinct of curiosity. Anyone whose imagination cannot be fired by the prospect of finding out something never before found by man will only waste his and others' time by taking up research, for only those will succeed who have a genuine interest and enthusiasm for discovery. The most successful scientists are capable of the zeal of the fanatic but are disciplined by objective judgment of their results and by the need to meet criticism from others. Love of science is likely to be accompanied by scientific taste and also is necessary to enable one to persist in the face of frustration.

A good intelligence, internal drive, willingness to work hard and tenacity of purpose are further prerequisites for success in research, as in nearly all walks of life. The scientist also needs imagination so that he can picture in his mind how processes work, how things take place that cannot be observed and conjure up hypotheses. The research worker is sometimes a difficult person because he has no great confidence in his opinions, yet he also is sceptical of others' views. This characteristic can be inconvenient in everyday life. Cajal commenting on the importance of mental independence in the scientist, remarks that humility may be fitting for saints but seldom for scientists.[110]

A spirit of indomitable perseverance has characterised nearly all successful scientists, for most worth while achievements required persistence and courage in face of repeated frustrations. So strong was this trait in Darwin that his son said it went beyond ordinary perseverance and could better be described as doggedness. Pasteur said :

> " Let me tell you the secret that has led me to my goal. My only strength lies in my tenacity."[112]

People may be divided roughly into those who habitually react vigorously to external influences—including ideas—and those who are passive and accept things as they come. The former question everything they are told even as children and often rebel against the conventional. They are curious and want to find out things for themselves. The other type fits into life with less trouble and, other things being equal, more easily accumulates information given as formal teaching. The mind of this latter type becomes furnished with generally accepted ideas and set opinions, whereas the reactive type has fewer fixed opinions and his mind remains free and flexible. Of course, not everyone can be classed as belonging to one of these two extremes, but clearly those approximating to the passive type are not cut out for research.

Preparing a list of the required attributes is not much help in the vexing problem of how to select promising people for research or of deciding yourself if you are suitable, because there is at present no objective means of measuring the qualities listed. However, this is a problem which psychologists might be able to solve in time. For example, it might be possible to devise a test of a person's knowledge of everyday things that would be a measure of his curiosity and powers of observation—his success in " discovering " things in his environment, for life can be a perpetual process of discovering. Tests might also be devised to measure ability to generalise, to formulate hypotheses to fit given data. Possibly love of science might be tested by determining the response—being delighted or not—on learning of scientific discoveries.

Ordinary examinations are not a good guide to a student's ability at research, because they tend to favour the accumulators of knowledge rather than the thinkers. Brilliant examinees are sometimes no good at research, while on the other hand some famous scientists have made a poor showing at examinations. Paul Ehrlich only got through his final medical examinations by the grace of the examiners who had the good sense to give recognition to his special talents, and Einstein failed at the entrance examination to the Polytechnic School. Probably the student who is reflective and critical is at a disadvantage in accumulating information as compared with the student who accepts without question all he is told. Charles Nicolle goes so far as to say that the inventive genius is not able to store knowledge and that inventiveness may be killed by bad teaching, fixed ideas and erudition.[63]

I have noticed that in England a great many research workers in both the biological and non-biological sciences are, or have been in their youth, keen naturalists. The pursuing of some branch of natural history as a hobby by a young man may be a valuable indication of an aptitude for research. It shows that he gets pleasure from studying natural phenomena and is curious to find out things for himself by observation.

At present the only way of selecting promising research talent —of " discovering discoverers " as Rous has put it—is by giving the candidate an opportunity of trying his hand at research for at least one or two years. Until the young scientist has shown that he has definite ability in research, it is wiser for him not to be given a permanent research position. This precaution is as important for the future welfare and happiness of the scientist as it is for the good of the research institution. It is helpful for undergraduates to be given an opportunity during their final year to dabble in research, as this often gives a preliminary indication of a person's suitability for research. One favourable indication is for the young graduate to show real desire to do research by taking steps to get a research position; in other words, the best research workers tend to select themselves.

Whatever the exact mental requirements may be, it is a

widely held opinion that not everyone is able to undertake research successfully, just as not everyone has talent for composing music, but lack of the particular requirements should not be regarded as a slur on the person's intelligence or his ability in other directions.

Incentives and rewards

The chief incentives of research are to satisfy curiosity, to satisfy the creative instinct, the desire to know whether one's conjecture has led to the creation of new knowledge and the desire for the feeling of importance by gaining recognition. More mundane incentives are the need to gain a livelihood and the ambition to "get on in the world", "showing" certain individuals who did not believe in your ability on the one hand, and on the other hand, trying to justify the confidence that others may have shown in you. Recognition of work done is an important incentive as is illustrated by the ill-feeling sometimes displayed over contentious points of priority in publication. Even great scientists are usually jealous of getting all due credit for their discoveries. The desire to see one's name in print and be credited throughout the scientific world with one's accomplishments is undoubtedly one of the most important incentives in research. In addition to these incentives which are common to all types of research, in applied research there is the desire to accomplish something for the good of mankind. This is likely to be more effective if it is not merely a vague ideal but if those to benefit are known to, or in some way associated with, the research worker.

The man or woman with a research mind is fascinated by the mental challenge of the unexplained and delights in exercising the wits in trying to find a solution. This is just a manifestation of the phenomenon that many people find pleasure in solving problems, even when there is no reward attached, as is shown by the popularity of crossword puzzles and detective stories. Incidentally Paul Ehrlich loved detective mysteries. Interest in a particular branch of science sometimes originates from the intrinsic beauty of the material or technique employed. Naturalists and zoologists are often attracted to study a group of animals because they find their appearance pleasing and a bacteriologist may like using a certain technique because it appeals to his aesthetic sensibility. Very likely it was Ehrlich's extraordinary love of bright colours (he is said to have derived an ecstatic pleasure from them) that gave him an interest in dyes and so determined the direction in which his work developed.

Albert Einstein distinguishes three types of research workers: those who take up science because it offers them an opportunity to exercise their particular talents and who exult in it as an athlete enjoys exercising his prowess; those who regard it as a means of livelihood and who but for circumstances might have become successful business men; and lastly the true devotees, who are rare but make a contribution to knowledge out of proportion to their numbers.[35]

Some psychologists consider that man's best work is usually done under adversity and that mental stress and even physical pain may act as a mental stimulant. Many prominent men have suffered from psychological troubles and various difficulties but for which perhaps they would never have put forward that effort required to excel.

The scientist seldom gets a large monetary reward for his labours so he should be freely granted any just fame arising from his work. But the greatest reward is the thrill of discovery. As many scientists attest, it is one of the greatest joys that life has to offer. It gives a tremendous emotional uplift and great sense of well-being and satisfaction. Not only factual discoveries but the sudden realisation of a generalisation can give the same feeling of exhilaration. As Prince Kropotkin wrote:

" He who has once in his life experienced this joy of scientific creation will never forget it."

Baker quotes the story of the great British biologist Alfred Wallace making a very small discovery:

" None but a naturalist," wrote Wallace, " can understand the intense excitement I experienced when at last I captured it [a new species of butterfly]. My heart began to beat violently, the blood rushed to my head, and I felt much more like fainting than I have done when in apprehension of immediate death. I had a headache the rest of the day, so great was the excitement produced by what will appear to most people a very inadequate cause."[8]

Referring to the elation he felt after demonstrating the feasibility of protecting people against smallpox by vaccination, Edward Jenner wrote:

> "The joy I felt at the prospect before me of being the instrument destined to take away from the world one of its greatest calamities . . . was so excessive that I sometimes found myself in a kind of reverie." [30]

Louis Pasteur and Claude Bernard made the following comments on this phenomenon:

> "When you have at last arrived at certainty, your joy is one of the greatest that can be felt by a human soul." [97]
> "The joy of discovery is certainly the liveliest that the mind of man can ever feel." [15]

The discoverer has an urge to share his joy with his colleagues and usually rushes into a friend's laboratory to recount the event and have him come and see the results. Most people get more fun and enjoyment out of new developments if they are able to share them with colleagues who are working on the same subject or are sufficiently closely related to be genuinely interested.

The stimulus of a discovery immediately wipes out all the disappointments of past frustrations and the scientist works with a new-found vigour. Furthermore, some stimulus is felt by his colleagues and so one discovery makes the conditions more propitious for further advances. But unfortunately things do not always turn out like this. Only too often our joy is short-lived and found to be premature. The consequent depression may be deep, and here a colleague can help by showing understanding and encouragement. To "take it" in this way without being beaten is one of the hard lessons the young scientist has to learn.

Unfortunately research has more frustrations than successes and the scientist is more often up against what appears to be an impenetrable barrier than making progress. Only those who have sought know how rare and hard to find are those little diamonds of truth which, when mined and polished, will endure hard and bright. Lord Kelvin wrote:

> "One word characterises the most strenuous of the efforts for the advancement of science that I have made perseveringly during fifty-five years; that word is failure."

Michael Faraday said that in the most successful instances less than one in ten of the hopes and preliminary conclusions are realised. When one is depressed, some cold comfort might be derived from the experience of those two great scientists. It is well for the young scientist to realise early that the fruits of research are not easily won and that if he is to succeed he will need endurance and courage.

The ethics of research

There are certain ethical considerations which are generally recognised among scientists. One of the most important is that, in reporting an investigation, the author is under an obligation to give due credit to previous work which he has drawn upon and to anyone who has assisted materially in the investigation. This elementary unwritten rule is not always followed as scrupulously as it should be and offenders ought to realise that increased credit in the eyes of the less informed readers is more than offset by the opprobrium accorded them by the few who know and whose opinion really matters. A common minor infringement that one hears is someone quoting another's ideas in conversation as though they were his own.

A serious scientific sin is to steal someone's ideas or preliminary results given in the course of conversation and to work on them and report them without obtaining permission to do so. This is rightly regarded as little better than common thieving and I have heard a repeated offender referred to as a " scientific bandit ". He who transgresses in this way is not likely to be trusted again. Another improper practice which unfortunately is not as rare as one might expect, is for a director of research to annex most of the credit for work which he has only supervised by publishing it under joint authorship with his own name first. The author whose name is placed first is referred to as the senior author, but senior in this phrase means the person who was responsible for most of the work, and not he who is senior by virtue of the post he holds. Most directors are more interested in encouraging their junior workers than in getting credit themselves. I do not wish to infer that in cases where the superior officer has played a real part in the work he should withhold his name altogether, as over-conscientious and generous people sometimes do, but often it is best to put it after that of the younger scientist so that the latter will not be overlooked as merely one of "and collaborators". The inclusion of the name of a well known scientist who has helped in the work is often useful as a guarantee of the quality of the work when the junior author has not yet established a reputation for himself. It is the duty of every scientist to give generously whatever advice and ideas he can and usually formal acknowledgment should not be demanded for such help.

Some colleagues and myself have found that sometimes what we have thought to be a new idea turns out not to be original at all when we refer to notes which we ourselves made on the subject some time previously. Incomplete remembering of this type occasionally results in the quite unintentional annexing of another person's idea. An idea given by someone else in conversation may subsequently be recalled without its origin being remembered and thus be thought to be one's own.

Complete honesty is of course imperative in scientific work. As Cramer said,

> "In the long run it pays the scientist to be honest, not only by not making false statements, but by giving full expression to facts that are opposed to his views. Moral slovenliness is visited with far severer penalties in the scientific than in the business world."[26]

It is useless presenting one's evidence in the most favourable light, for the hard facts are sure to be revealed later by other investigators. The experimenter has the best idea of the possible errors in his work. He should report sincerely what he has done and, when necessary, indicate where mistakes may have arisen.

If an author finds out he cannot later substantiate some results he has reported he should publish a correction to save others either being misled or put to the trouble of repeating the work themselves, only to learn that a mistake has been made.

When a new field of work is opened up by a scientist, some people consider it courteous not to rush in to it, but to leave the field to the originator for a while so that he may have an opportunity of reaping the first fruits. Personally I do not see any need to hold back once the first paper has been published.

Hardly any discovery is possible without making use of a knowledge gained by others. The vast store of scientific knowledge which is to-day available could never have been built up if scientists did not pool their contributions. The publication of experimental results and observations so that they are available to others and open to criticism is one of the fundamental principles on which modern science is based. Secrecy is contrary to the best interests and spirit of science. It prevents the individual contributing to further progress; it usually means that he or his employer is trying to exploit for their own gain some advance made by building on the knowledge which others have freely given. Much research is carried out in secret in industry and in government war departments. This seems to be inevitable in the world as it is to-day, but it is nevertheless wrong in principle. Ideally, freedom to publish, provided only that the work has sufficient merit, should be a basic right of all research workers. It is said that occasionally, even in agricultural research, results may be suppressed because they are embarrassing to government authorities.[54] This would seem to be a dangerous and shortsighted policy.

Personal secrecy in laboratories not subject to any restrictions is not infrequently shown by workers who are afraid that someone else will steal their preliminary results and bring them to fruition and publish before they themselves are able to do so. This form of temporary secrecy can hardly be regarded as a breach of scientific ethics but, although understandable, it is not commendable, for free interchange of information and ideas helps hasten the advance of science. Nevertheless information given in confidence must be respected as such and not handed on to others. A travelling scientist visiting various laboratories may himself be perfectly honourable in not taking advantage of unpublished information he is given, but may inadvertently hand on such information to a less scrupulous individual. The traveller can best avoid this risk by asking not to be told anything that is wished to be kept confidential, for it is difficult to remember what is for restricted distribution and what not.

Even in the scientific world, unfortunately, one occasionally encounters national jealousies. These are manifest by lack of appreciation or acknowledgment of work done in other countries.

Not only is this to be deplored as a quite indefensible breach of ethics and of the international spirit of science, but it rebounds on the offenders, often to the detriment of themselves and their country. The person failing to appreciate advances in science made elsewhere may be left in the backwater he deserves, and he shows himself a second-rate scientist. Among the great majority of scientists there exists an international freemasonry that is one of the main reasons for faith in the future of mankind, and it is depressing to see this marred by petty selfishness on the part of a few individuals.

Different types of scientific minds

Not all minds work alike. Attempts are often made to divide scientists broadly into two types, but the classification is arbitrary and probably the majority fall somewhere between the two extremes and combine many of the characteristics of both.

W. D. Bancroft,[10] the American chemist, calls one type the "guessers" (using the word guess in the sense of making a shrewd judgment or hypothesis in advance of the facts): these follow mainly the deductive or Aristotelian methods. They get their hypothesis first, or at any rate early in the investigation, and then test it by experiment. The other type he calls the "accumulators" because they accumulate data until the generalisation or hypothesis is obvious; these follow the inductive or Baconian method. However, the terms inductive and deductive, and Aristotelian and Baconian can be confusing and have sometimes been misused. Henri Poincaré[72] and Jacques Hadamard[50] classify mathematicians as either "intuitive" or "logical" according to whether they work largely by intuitions or by gradual systematic steps. This basis of classification seems to agree with Bancroft's. I will use the terminology "speculative" and "systematic" as this seems the simplest way of indicating the principal difference between the two types.

Charles Nicolle[63] distinguished (a) the inventive genius who cannot be a storehouse for knowledge and who is not necessarily highly intelligent in the usual sense, and (b) the scientist with a fine intelligence who classifies, reasons and deduces but is, according to Nicolle, incapable of creative originality or making original discoveries. The former uses intuition and only calls on logic and reason to confirm the finding. The latter advances knowledge by gradual steps like a mason putting brick on brick until finally a structure is formed. Nicolle says that intuitions were so strong with Pasteur and Metchnikoff that sometimes they almost published before the experimental results were obtained. Their experiments were done mainly to reply to their critics.

Bancroft gives the following illustrations of the outlook of the different types of scientist. Examples of the systematic type are Kelvin and Sir W. Hamilton, who said,

" Accurate and minute measurement seems to the non-scientific imagination a less lofty and dignified work than looking for something new, yet nearly all the grandest discoveries are made this way ",

" In physical sciences the discovery of new facts is open to any blockhead with patience and manual dexterity and acute senses."

Contrast this last statement with one made by Davy :

" I thank God I was not made a dextrous manipulator; the most important of my discoveries have been suggested to me by my failures."

Most mathematicians are the speculative type. The following remarks are attributed to Newton, Whewell and Gauss respectively :

" No great discovery is ever made without a bold guess,"

" Advances in knowledge are not commonly made without some boldness and licence in guessing,"

" I have the result but I do not yet know how to get it."

Most of the outstanding discoverers in biology have also been of the speculative type. Huxley wrote :

" It is a popular delusion that the scientific enquirer is under an obligation not to go beyond generalisation of observed facts ... but anyone who is practically acquainted with scientific work is aware that those who refuse to go beyond the facts, rarely get as far."

The following two comments, made on different occasions, reveal Pasteur's views on this point :

" If someone tells me that in making these conclusions I have gone beyond the facts, I reply: 'it is true that I have freely put myself among ideas which cannot be rigorously proved. That is my way of looking at things.' "

" Only theory can bring forth and develop the spirit of invention."

W. Ostwald classifies scientists slightly differently.[67] He distinguishes the classicist whose main characteristic is to bring to perfection every discovery and is systematic, and the romanticist who has a multitude of ideas but has a certain amount of superficiality in dealing with them and seldom works them out completely. Ostwald says the classicist is a bad teacher and cannot do anything in front of others, while the romanticist gives away his ideas freely and has an enormous influence on his students. He may produce some outstanding students but sometimes spoils their originality. On the other hand, as Hadamard points out, highly intuitive minds may be very obscure. Kenneth Mees considers that practical scientific discovery and technology embrace three different methods of working: (a) theoretical synthesis, (b) observation and experiment, (c) invention. It is rare, he says, for one man to excel in more than one of these activities, for each requires a different type of mind.[61]

The systematic type of scientist is probably more suited to developmental research and the speculative type to exploratory research; the former to team work and the latter either to individual work or as leader in a team. Dr. E. L. Taylor describes the organisation of a large commercial research organisation which employed men of the speculative type to play about with their ideas, but as soon as they hit on something that promised to be of value it was taken out of their hands entirely and given to a systematic worker to test and develop fully.[90]

The speculative and systematic types, however, represent extremes and probably most scientists combine some of the characteristics of both. The student may find that he has natural tendencies toward one type or the other. Bancroft considers that often one type cannot be converted to the other. It is probably best for each to follow his natural tendencies and one wonders if many scientists have not been unduly influenced by the teacher under whose influence they happened to fall. The important thing is for us not to expect everyone to think the same way as we do ourselves. It is a great pity for a young scientist who is naturally the speculative type to come under the influence of a systematic type and be misguided into believing that his imagination should be suppressed to the extent that it is crushed. The man who gets ideas of his own and wants to try them out is more likely to be attracted by research, to contribute more to it, and to get more from it than the man lacking in imagination and curiosity. The latter can do useful work on research but probably does not get much enjoyment out of it. Both types are necessary for the advancement of science for they tend to be complementary.

As is mentioned elsewhere, it is a common error among philosophers and writers of books on the scientific method to believe that discoveries are made by the systematic accumulation of data until the generalisation is a matter of plain logic, whereas in fact this is true in probably a minority of cases.

The scientific life

Some comment on the personal aspects of research might be helpful to the young man or woman contemplating taking up a scientific career.

The young scientist on reading this book might be alarmed at the demands made on him and, unless he is one of those rare individuals who is willing to give his whole life to "a cause", he may be put off research if some further comment is not offered. Let me reassure him at once that this is a counsel of perfection and one can become a good research worker without sacrificing all other interests in life. If one is willing to regard research as a calling and to become what Einstein calls a true devotee, all to the good, but there are plenty of examples of great and successful scientists who have not only lived normal family lives but have managed also to find time for many outside interests. Until recent times research was carried on only by the devotees, because the material rewards were so poor, but nowadays research has become a regular profession. However, it cannot be conducted successfully on a strict 9 a.m. to 5 p.m. basis and some evening study is a practical necessity. One needs to have a real interest in science and it must be part of one's life and looked upon as a pleasure and a hobby.

Research work progresses in an irregular manner and only occasionally is the scientist hotly pursuing a new discovery. It is then he needs to pour all his energies into the work and think of it day and night. If he has the true scientific spirit he will want to do this and it is crippling if circumstances prevent it. The research man's family usually understand that if he is to be a creative scientist, there are times when it is most important for him to be spared other responsibilities and worries as much as possible; and likewise his colleagues at the laboratory usually try and help with any other commitments he may have in the way of routine work or administration. This help is not likely to be a burden on his associates or family because these spurts are all too rare with most people. Perhaps two to six intervals each of a week or two every year might be average, but they will vary enormously from one individual to another. However, these remarks should not be misconstrued as an encouragement to develop an "artistic temperament" and lack of responsibility in everyday affairs!

When Simon Flexner was planning the Rockefeller Institute he was asked "are you going to allow your men to make fools of themselves at your Institute?" The implication was that only those who would risk doing so were likely to make important discoveries. The research man must not be put off his ideas by fear of being ridiculous or being said to have "a bee in his bonnet". It sometimes requires courage to put forward and follow up a novel idea. It will be remembered that Jenner confided his proposals about vaccination to a friend under a bond of secrecy for fear of ridicule.

When I asked Sir Alexander Fleming about his views on research his reply was that he was not doing research when he discovered penicillin, he was just playing. This attitude is typical of many bacteriologists who refer to their research as "playing about" with this or that organism. Sir Alexander believes that it is the people who play about who make the initial discoveries and the more systematic scientists who develop them. This expression, "playing about", is significant for it clearly means that the scientist is doing something for his own enjoyment, to satisfy his own curiosity. However, with the incompetent person "playing about" may amount to nothing more than ineffectual pottering in which nothing is followed up. Sir Henry Dale, speaking at a Congress held in Cambridge in 1948 in honour of Sir Joseph Barcroft, said that the great physiologist always regarded research as an amusing adventure. Speaking at the same Congress, Professor F. J. W. Roughton said that for Barcroft and for Starling, physiology was the greatest sport in the world.

The great pioneers of science, although they have defended their ideas fervently and often fought for them, were mostly at heart humble men, for they realised only too clearly how puny were their achievements compared to the vastness of the as yet unknown. Near the end of his life Pasteur said : "I have wasted my life" as he thought of the things he might have done to greater profit. Shortly before his death Newton is reported to have said :

> "I know not what I may appear to the world, but to myself I appear to have been only like a boy playing on the sea-shore, and diverting myself in now and then finding a smoother pebble or a prettier shell than ordinary, whilst the great ocean of truth lay all undiscovered before me."

Diversion and holidays are very much a question of individual requirements but freshness and originality may be lost if the scientist works unremittingly for too long. In this connection a good maxim has been coined by Jowett: "Don't spare; don't drudge." Most of us require recreation and variety in interests to avoid becoming dull, stodgy and mentally constipated. Simon Flexner's attitude to holidays was the same as Pierpont Morgan's —who once remarked that he could do a full year's work in nine months but not in twelve months. Most scientists, however, do not require as much as three months' annual vacation.

Mention has already been made of the disappointments so often met in research and the need for understanding and encouragement from colleagues and friends. It is recognised that these continual frustrations sometimes produce a form of neurosis which Professor H. A. Harris calls "lab. neurosis", or they may kill a man's interest in research. Interest and enthusiasm must be kept alive and this may be difficult if the worker is obliged to plod along on a line of work which is not getting anywhere.

In most walks of life it is possible to get into a groove, or to go "stale", but it is a more serious problem in research than in most other occupations, because practically all the research worker's activities must be initiated from within his own brain. He gets stimulus from his work only when he is making progress, whereas the business man, the lawyer and the physician are constantly receiving stimulus both from their clients and from the fact that they are able to effect something.

Frequent discussion of one's work with associates who show an interest in it is helpful in avoiding "lab. neurosis". The great value of "mental catharsis" in neurosis is well known, and similarly telling others of one's problems and sharing one's disappointments can help the baffled research worker from suffering unduly from worry.

"Lab. neurosis" is most likely to arise in scientists devoting all their time to one research problem. Some individuals find sufficient relief if they have two problems under investigation at the same time. For others it is better to spend a portion of their time in teaching, routine diagnostic work, administration or similar occupation which enables them to feel they are doing something effectively and contributing something to the community even if getting nowhere with the research. Each case needs to be considered individually, but if effective research is to be accomplished the scientist nevertheless has to devote the major portion of his time to it.

With regard to this latter point W. B. Cannon waxes eloquent:

> "This time element is essential. The investigator may be made to dwell in a garret, he may be forced to live on crusts and wear dilapidated clothes, he may be deprived of social recognition, but if he has time, he can steadfastly devote himself to research. Take away his free time and he is utterly destroyed as a contributor to knowledge."[22]

It is little use to squeeze research into an hour or two of spare time during a day occupied in other duties, especially if the other duties are of a nature that require a lot of thought, for, apart from time at the bench, research requires peace of mind for reflection. Furthermore, to achieve results in research it is sometimes necessary to drive oneself in the face of frustrations and it may be a disadvantage to have a too ready alternative "escape" activity. F. M. Burnet considers that part-time research is usually "of relatively unimportant character".

Platt and Baker suggest that a research worker may have to choose between having a reputation as being good natured and easily accessible to visitors but mediocre, or on the other hand temperamental but successful. Visitors to laboratories who are merely scientific sightseers ought to be severely discouraged, but most research workers are glad to make time to talk to visitors who have a genuine and serious interest in their work.

Just before his death Pavlov wrote:

> "What can I wish to the youth of my country who devote themselves to science? *Firstly, gradualness.* About this most important condition of fruitful scientific work I can never speak without emotion. Gradualness, gradualness, gradualness . . . never begin the subsequent without mastering the preceding . . . But do not become the archivist of facts. Try to penetrate the secret of their occurrence, persistently searching for the laws which govern them. *Secondly, modesty* . . . do not allow haughtiness to take you in possession. Due to that you will be obstinate where it is necessary to agree, you will refuse useful and friendly help, you will lose your objectiveness. *Thirdly, passion.* Remember that science demands from a man all his life. If you had two lives that would not be enough for you. Be passionate in your work and your searching."[68]

Enthusiasm is one of the great motivating forces, but, like anything associated with emotion, it can be fickle. Some people are given to bursts of intense enthusiasm which are short-lived, whereas others are able to sustain their interest for long periods, usually at a more moderate intensity. It is as well to learn as much as possible about oneself in this, as in other respects. Personally, when I feel myself in the grip of an enthusiasm, warned by past experience, I try to assess the situation objectively and decide if there is a solid foundation for the enthusiasm or if it is likely to burn itself out leaving that deflated feeling from which it is difficult to rouse further interest in the subject. One help in sustaining interest in a subject is to share that interest with colleagues. This also helps to sober one up and check ill-founded bursts of enthusiasm. Young people are especially liable to get excited about their ideas and be impatient to try them out without giving them sufficient critical thought. Enthusiasm is a most valuable stimulant but, like most stimulants, its use needs to be tempered with a proper understanding of all its effects.

If the young scientist succeeds within a year or two of graduation in establishing a profitable line of work, it is as well for him to pursue it to the exclusion of other subjects, but generally it is wise for him to gain some breadth of experience before devoting all his time to one field. Similarly with his place of work : if he is fortunate enough to find his colleagues and the circumstances of his position such that he is well satisfied with his advances, well and good, but often, especially if the scientist feels he is getting into a groove, a change of position is very helpful owing to the great stimulus that is to be had from fresh mental contacts and different scientific fields. I have been struck by this myself and others have told me that they also have experienced it. Perhaps every three to five years the scientist under forty should examine his position in this light. A change of subjects also is often beneficial, for working too long on the same problem can produce intellectual sterility.

It is usually difficult or undesirable for senior scientists to change their posts; for them the sabbatical year's leave provides the opportunity for a change of mental climate, while another method is to arrange a temporary exchange of scientists between institutes.

It is rare for a person to carry within himself enough drive and interest to be able to pursue research for long if he is isolated from people with similar interests. Most scientists stagnate when alone, but in a group have a symbiotic-like effect on one another, just as to culture some bacteria it is necessary to have a number of individual organisms or to start a fire several sticks are necessary. This is the main advantage of working in a research centre. The fact that there one can get advice and co-operation from colleagues and borrow apparatus is of secondary importance. Scientists from the more outlying parts of the world get great benefit from coming to one of the great research centres for a period of work, and also from paying brief visits to various research centres. Similarly, the main value of scientific congresses is the opportunity they provide for scientists to meet informally and discuss topics of mutual interest. Great stimulus is to be derived from meeting people who are interested in the same things as ourselves, and subjects become more interesting when we see how interested others are in them. Indeed few of us are sufficiently strong-minded and independent to be enthusiastic about a subject which does not interest others.

Nevertheless there are the rare individuals who have sufficient internal drive and enthusiasm not to stagnate when alone and even perhaps to benefit from the forced independence and wider interests that the isolated worker is obliged to take up. Most of the great pioneers had to work out their ideas independently and some—Mendel in his monastery and Darwin during the voyage of the *Beagle*—worked in scientific isolation. A present-day example is H. W. Bennetts who has worked in comparative scientific isolation in Western Australia. He has to his credit the discovery of the cause of entero-toxaemia of sheep and copper deficiency as a cause of disease in sheep and cattle as well as other important pioneer contributions to knowledge.

Lehman has collected some interesting data about man's most creative time of life.[59] He extracted data from sources such as *A Series of Primers of the History of Medicine* and *An Introduction to the History of Medicine*, and found that the maximum output of people born between 1750 and 1850 was during the decade of life 30 to 39 years. Taking this as 100 per cent, the output for the decade 20–29 years was 30–40 per cent; for 40–49 years, 75 per cent; 50–59 years, about 30 per cent. Probably man's inventiveness and originality begins to decrease at an early age, possibly even in the 20s, but this is offset by increased experience, knowledge and wisdom.

Cannon says that Long and Morton began the use of ether as an anaesthetic when they were both 27 years of age; Banting was 31 when he discovered insulin; Semmelweis recognised the infectiousness of puerperal fever when he was 29; Claude Bernard had started his researches on the glycogenic function of the liver when he was 30; van Grafe devised the operation for cleft palate and founded modern plastic surgery when he was 29. When von Helmholtz was only 22, barely emerged as an undergraduate medical student, he published an important paper suggesting that fermentation and putrefaction were vital phenomena and thus paved the way for Pasteur.[58] Robinson considers 28 is a critical age, as many great scientists have published their most important work at that age. On the other hand, some individuals continue to do first-rate research till they are past 70. Pavlov, Sir Frederic Gowland Hopkins and Sir Joseph Barcroft are good examples.

The fact that a person has not made a significant contribution by the time he is 40 does not necessarily mean he never will, for such cases have occurred, though not often. With advancing age most minds become less receptive to new ideas suggested by others and probably also arising from their work or thinking. William Harvey stated that no man over forty accepted the idea of the circulation when he first advanced it. The reason why many lose their productivity about middle age is often simply due to their having taken on administrative responsibilities that do not allow time for research. In other cases indolence develops with middle age and security, and drive is lost. Contact with young minds often helps to preserve freshness of outlook. Whatever the reasons for the frequent falling off of productivity after middle age, its occurrence shows that accumulation of knowledge and experience is not the main factor in successful research.

W. Ostwald considered that the frequent decrease of productivity with increasing age is due to too long familiarity with the same subject. The way in which accumulated information handicaps originality was discussed in the first chapter of this book. For scientists past middle age who have lost originality, Ostwald advocated a radical change of field of work. In his own case he was evidently successful in refreshing his mind by this means when he was over fifty years of age.

The research scientist is fortunate in that in his work he can find something to give meaning and satisfaction to life. For those who seek peace of mind by sinking their personality in something bigger than themselves, science can have a special appeal, while the somewhat more material-minded can get gratification from the knowledge that his achievements in research have an immortality. Few callings can claim to have as much influence on the welfare of mankind as scientific research, especially in the medical and biological sciences. Brailsford Robertson said : " The investigator is the pathfinder and the pioneer of new civilisations."[4] The human race has existed and been accumulating knowledge for only about a million years, and civilisation started only some 10,000 years ago. There is no known reason why the world should not remain habitable for hundreds of millions of years to come. The mind staggers at the thought of what will be accomplished in the future. We have scarcely begun to master the forces of nature.

But more urgent than finding out how to control the world's climate, to draw on the heat stored under the crust of the earth, or reaching out through space to other worlds, is the need for man's social development to catch up with his achievements in the physical sciences. And whose fancy can guess at the shape of things to come when mankind finds the collective will and courage to assume the tremendous but ultimately inescapable responsibility of deliberately directing the further evolution of the human species, and the greatest tool of research, the mind of man, becomes itself the subject of scientific development?

SUMMARY

Curiosity and love of science are the most important mental requirements for research. Perhaps the main incentive is the desire to win the esteem of one's associates, and the chief reward is the thrill of discovery, which is widely acclaimed as one of the greatest pleasures life has to offer.

Scientists may be divided broadly into two types according to their method of thinking. At one extreme is the speculative worker whose method is to try to arrive at the solution by use of imagination and intuition and then test his hypothesis by experiment or observation. The other extreme is the systematic worker who progresses slowly by carefully reasoned stages and who collects most of the data before arriving at the solution.

Research work commonly progresses in spurts. It is during the "high spots" that it is almost essential for the scientist to devote all possible energy and time to the work. Continual frustrations may produce a mild form of neurosis. Precautions against this include working on more than one problem at a time or having some other part-time occupation. A change of mental environment usually provides a great mental stimulus, and sometimes a change of subject does too.

There is real gratification to be had from the pursuit of science, for its ideals can give purpose to life.

APPENDIX

(1) It was not a physicist but a physiologist, Luigi Galvani, who discovered current electricity. He had dissected a frog and left it on a table near an electrical machine. When Galvani left it for a moment someone else touched the nerves of the leg with a scalpel and noticed this caused the leg muscles to contract. A third person noticed that the action was excited when there was a spark from the electric machine. When Galvani's attention was drawn to this strange phenomenon he excitedly investigated it and followed it up to discover current electricity.[112]

(2) In 1822 the Danish physicist, Oersted, at the end of a lecture happened to bring a wire, joined at its two extremities to a voltaic cell, to a position above and parallel to a magnetic needle. At first he had purposely held the wire perpendicular to the needle but nothing happened, but when by chance he held the wire horizontally and parallel to the needle he was astonished to see the needle change position. With quick insight he reversed the current and found that the needle deviated in the opposite direction. Thus by mere chance the relationship between electricity and magnetism was discovered and the path opened for the invention by Faraday of the electric dynamo. It was when telling of this that Pasteur made his famous remark : " In the field of observation chance favours only the prepared mind." Modern civilisation perhaps owes more to the discovery of electro-magnetic induction than to any other single discovery.[69]

(3) When von Röntgen discovered X-rays he was experimenting with electrical discharges in high vacua and using barium platinocyanide with the object of detecting invisible rays, but had no thought of such rays being able to penetrate opaque materials. Quite by chance he noticed that barium platino-

cyanide left on the bench near his vacuum tube became fluorescent although separated from the tube by black paper. He afterwards said: " I found by accident that the rays penetrated black paper."[8]

(4) When W. H. Perkin was only eighteen years old he tried to produce quinine by the oxidation of allyl-o-toluidine by potassium dichromate. He failed, but thought it might be interesting to see what happened when a simpler base was treated with the same oxidiser. He chose aniline sulphate and thus produced the first aniline dye. But chance played an even bigger part than the bare facts indicate: had not his aniline contained as an impurity some p-toluidine the reaction could not have occurred.[8]

(5) During the first half of the nineteenth century it was firmly believed that animals were unable to manufacture carbohydrates, fats or proteins, all of which had to be obtained in the diet preformed from plants. All organic compounds were believed to be synthesised in plants whereas animals were thought to be capable only of breaking them down. Claude Bernard set out to investigate the metabolism of sugar and in particular to find where it is broken down. He fed a dog a diet rich in sugar and then examined the blood leaving the liver to see if the sugar had been broken down in the liver. He found a high sugar content, and then wisely carried out a similar estimation with a dog fed a sugar-free meal. To his astonishment he found also a high sugar content in the control animal's hepatic blood. He realised that contrary to all prevailing views the liver probably did produce sugar from something which is not sugar. Thereupon he set about an exhaustive series of experiments which firmly established the glycogenic activity of the liver. This discovery was due firstly to the fact that Bernard was meticulous in controlling every stage of his experiments, and secondly, to his ability to recognise the importance of a result discordant with prevailing ideas on the subject and to follow up the clue thus given.[44]

(6) A mixture of lime and copper sulphate was sprayed on posts supporting grape vines in Medoc with the object of frightening away pilferers. Millardet later noticed that leaves accidentally sprayed with the mixture were free from mildew.

The following up of this clue led to the important discovery of the value of Bordeaux mixture in protecting fruit trees and vines from many diseases caused by fungi.[87]

(7) The property of formalin of removing the toxicity of toxins without affecting their antigenicity was discovered by Ramon by chance when he was adding antiseptics to filtrates with the object of preserving them.[63]

(8) The circumstances leading to the discovery of penicillin are widely known. Fleming was working with some plate cultures of staphylococci which he had occasion to open several times and, as often happens in such circumstances, they became contaminated. He noticed that the colonies of staphylococci around one particular colony died. Many bacteriologists would not have thought this particularly remarkable for it has long been known that some bacteria interfere with the growth of others. Fleming, however, saw the possible significance of the observation and followed it up to discover penicillin, although its development as a therapeutic agent was due to the subsequent work of Sir Howard Florey. The element of chance in this discovery is the more remarkable when one realises that that particular mould is not a very common one and, further, that subsequently a most extensive, world-wide search for other antibiotics has failed to date to discover anything else as good. It is of interest to note that the discovery would probably not have been made had not Fleming been working under "unfavourable" conditions in an old building where there was a lot of dust and contaminations were likely to occur.[42]

(9) J. Ungar[96] found that the action of penicillin on certain bacteria was slightly enhanced by the addition to the medium of paraminobenzoic acid (PABA). He did not explain what made him try this out but it seems likely that it was because PABA was known to be an essential growth factor for bacteria. Subsequently, Greiff, Pinkerton and Moragues[49] tested PABA to see if it enhanced the weak inhibitory effect which penicillin had against typhus rickettsiae. They found that PABA alone had a remarkably effective chemotherapeutic action against the typhus organisms. "This result was quite unexpected," they said. As a result of this work PABA became recognised as a valuable chemotherapeutic agent for the typhus

group of fevers, against which previously nothing had been found effective.

In the chapter on hypothesis I have described how salvarsan and sulphanilamide were discovered following an hypothesis that was not correct. Two other equally famous chemotherapeutic drugs were discovered only because they happened to be present as impurities in other substances which were being tested. Scientists closely associated with the work have told me the stories of these two discoveries but have asked me not to publish them as other members of the team may not wish the way in which they made the discovery to be made public. Sir Lionel Whitby has told to me a story of a slightly different nature. He was conducting an experiment on the then new drug, sulphapyridine, and mice inoculated with pneumococci were being dosed throughout the day, but were not treated during the night. Sir Lionel had been out to a dinner party and before returning home visited the laboratory to see how the mice were getting on, and while there lightheartedly gave the mice a further dose of the drug. These mice resisted the pneumococci better than any mice had ever done before. Not till about a week later did Sir Lionel realise that it was the extra dose at midnight which had been responsible for the excellent results. From that time, both mice and men were dosed day and night when under sulphonamide treatment and they benefited much more than under the old routine.

(10) In my researches on foot-rot in sheep I made numerous attempts to prepare a medium in which the infective agent would grow. Reason led me to use sheep serum in the medium and the results were repeatedly negative. Finally I got a positive result and on looking back over my notes I saw that, in that batch of media, horse serum had been used in place of sheep serum because the supply of the latter had temporarily run out. With this clue it was a straightforward matter to isolate and demonstrate the causal agent of the disease—an organism which grows in the presence of horse serum but not sheep serum! Chance led to a discovery where reason had pointed in the opposite direction.

(11) The discovery that the human influenza virus is able to infect ferrets was a landmark in the study of human respiratory

diseases. When an investigation on influenza was planned, ferrets were included among a long list of animals it was intended to try and infect sooner or later. However, some time before it was planned to try them, it was reported that a colony of ferrets was suffering from an illness which seemed to be the same as the influenza then affecting the people caring for them. Owing to this circumstantial evidence, ferrets were immediately tried and found susceptible to influenza. Afterwards it was found that the idea which prompted the tests in ferrets was quite mistaken for the disease occurring in the colony of ferrets was not influenza but distemper![3]

(12) A group of English bacteriologists developed an effective method of sterilising air by means of a mist made from a solution of hexyl-resorcinol in propylene-glycol. They conducted a very extensive investigation trying out many mixtures. This one proved the best; the glycol was chosen merely as a suitable vehicle for the disinfectant, hexyl-resorcinol. Considerable interest was aroused by the work because of the possibility of preventing the spread of air-borne diseases by these means. When other investigators took up the work they found that the effectiveness of the mixture was due not to the hexyl-resorcinol but to the glycol. Subsequently, glycols proved to be some of the best substances for air disinfection. They were only introduced into this work as solvents for other supposedly more active disinfectants and were not at first suspected as having any appreciable disinfective action themselves.[73]

(13) Experiments were being conducted at Rothamsted Experimental Station on protecting plants from insects with various compounds, when it was noticed that those plants treated with boric acid were strikingly superior to the rest. Investigation by Davidson and Warington showed that the better growth had resulted because the plants required boron. Previously it had not been known that boron was of any importance in plant nutrition and even after this discovery, boron deficiency was for a time thought of as only of academic interest. Later, however, some diseases of considerable economic importance—"heart-rot" of sugar beet for example—were found to be manifestation of boron deficiency.[102]

(14) The discovery of selective weed-killers arose unexpectedly

from studies on root nodule bacteria of clovers and plant growth stimulants. These beneficial bacterial nodules were found to exert their deforming action on the root hairs by secreting a certain substance. But when Nutman, Thornton and Quastel tested the action of this substance on various plants, they were surprised to find that it prevented germination and growth. Furthermore they found that this toxic effect was selective, being much greater against dicotyledon plants, which include most weeds, than against monocotyledon plants, which include grain crops and grasses. They then tried related compounds and found some which are of great value in agriculture to-day as selective weed-killers.[65]

(15) Scientists working on the technicalities of food preservation tried prolonging the "life" of chilled meat by replacing the air by carbon dioxide which was known to have an inhibitory effect on the growth of micro-organisms causing spoilage. Carbon dioxide, at the high concentration used, was found to cause an unpleasing discoloration of the meat and the whole idea was abandoned. Some time later, workers in the same laboratory were investigating a method of refrigeration which involved the release of carbon dioxide into the chamber in which the food was stored, and observations were carried out to see whether the gas had any undesirable effect. To their surprise the meat not only remained free from discoloration but even in the relatively low concentrations of carbon dioxide involved it kept in good condition much longer than ordinarily. From this observation was developed the important modern process of "gas storage" of meat in which 10–12 per cent carbon dioxide is used. At this concentration the gas effectively prolongs the "life" of chilled meat without causing discoloration.[13]

(16) I was investigating a disease of the genitalia of sheep known as balano-posthitis. It is a very long-lasting disease and was thought to be incurable except by radical surgery. Affected sheep were sent from the country to the laboratory for investigation but to my surprise they all healed spontaneously within a few days of arrival. At first it was thought that typical cases had not been sent, but further investigation showed that the self-imposed fasting of the sheep when placed in a strange environment had cured the disease. Thus it was found that

this disease, refractory to other forms of treatment, could in most cases be cured by the simple expedient of fasting for a few days.

(17) Paul Ehrlich's discovery of the acid-fast method of staining tubercle bacilli arose from his having left some preparations on a stove which was later inadvertently lighted by someone. The heat of the stove was just what was required to make these waxy-coated bacteria take the stain. Robert Koch said "We owe it to this circumstance alone that it has become a general custom to search for the bacillus in sputum."[113]

(18) Dr. A. S. Parkes relates the following story of how he and his colleagues made the important discovery that the presence of glycerol enables living cells to be preserved for long periods at very low temperatures.

"In the autumn of 1948 my colleagues. Dr. Audrey Smith and Mr. C. Polge, were attempting to repeat the results which Shaffner, Henderson and Card (1941) had obtained in the use of laevulose solutions to protect fowl spermatozoa against the effects of freezing and thawing. Small success attended the efforts, and pending inspiration a number of the solutions were put away in the cold-store. Some months later work was resumed with the same material and negative results were again obtained with all of the solutions except one which almost completely preserved motility in fowl spermatozoa frozen to –79°C. This very curious result suggested that chemical changes in the laevulose, possibly caused or assisted by the flourishing growth of mould which had taken place during storage, had produced a substance with surprising powers of protecting living cells against the effects of freezing and thawing. Tests, however, showed that the mysterious solution not only contained no unusual sugars, but in fact contained no sugar at all. Meanwhile, further biological tests had shown that not only was motility preserved after freezing and thawing but, also, to some extent, fertilizing power. At this point, with some trepidation, the small amount (10–15 ml.) of the miraculous solution remaining was handed over to our colleague Dr. D. Elliott for chemical analysis. He reported that the solution contained glycerol, water, and a fair amount of protein! It was then realised that Mayer's albumen—the glycerol and albumen of the histologist—had been used in the course of morphological work on the spermatozoa at the same time as the laevulose solutions were being tested, and with them had been put away in the cold-store. Obviously there had been some confusion with the various bottles, though we never found out exactly what had

166

happened. Tests with new material very soon showed that the albumen played no part in the protective effect, and our low temperature work became concentrated on the effects of glycerol in protecting living cells against the effects of low temperatures."[115]

(19) In a personal communication Dr. A. V. Nalbandov has given the following intriguing story of how he discovered the simple method of keeping experimental chickens alive after the surgical removal of the pituitary gland (hypophysectomy).

" In 1940 I became interested in the effects of hypophysectomy of chickens. After I had mastered the surgical technique my birds continued to die and within a few weeks after the operation none remained alive. Neither replacement therapy nor any other precautions taken helped and I was about ready to agree with A. S. Parkes and R. T. Hill who had done similar operations in England, that hypophysectomized chickens simply cannot live. I resigned myself to doing a few short-term experiments and dropping the whole project when suddenly 98% of a group of hypophysectomized birds survived for 3 weeks and a great many lived for as long as 6 months. The only explanation I could find was that my surgical technique had improved with practice. At about this time, and when I was ready to start a long-term experiment, the birds again started dying and within a week both recently operated birds and those which had lived for several months, were dead. This, of course, argued against surgical proficiency. I continued with the project since I now knew that they could live under some circumstances which, however, eluded me completely. At about this time I had a second successful period during which mortality was very low. But, despite careful analysis of records (the possibility of disease and many other factors were considered and eliminated) no explanation was apparent. You can imagine how frustrating it was to be unable to take advantage of something that was obviously having a profound effect on the ability of these animals to withstand the operation. Late one night I was driving home from a party via a road which passes the laboratory. Even though it was 2 A.M. lights were burning in the animal rooms. I thought that a careless student had left them on so I stopped to turn them off. A few nights later I noted again that lights had been left on all night. Upon enquiry it turned out that a substitute janitor, whose job it was to make sure at midnight that all the windows were closed and doors locked, preferred to leave on the lights in the animal room in order to be able to find the exit door (the light switches not being near the door). Further checking showed that the two survival periods coincided with the times when the substitute janitor was on the job. Controlled experiments soon showed that hypophysectomized chickens kept in darkness all died while chickens lighted for 2 one-hour periods nightly lived indefinitely. The explanation was that birds in the dark do not eat and develop hypoglycaemia from which they cannot recover, while birds which are lighted eat enough to prevent hypoglycaemia. Since that time we no longer experience any trouble in maintaining hypophysectomized birds for as long as we wish."

BIBLIOGRAPHY

1. Allbutt, C. T. (1905). *Notes on the Composition of Scientific Papers*. Macmillan & Co. Ltd., London.
2. Anderson, J. A. (1945). "The preparation of illustrations and tables." *Trans. Amer. Assoc. Cereal Chem.*, 3, 74.
3. Andrewes, C. H. (1948). Personal communication.
4. Annual Report, New Zealand Dept. Agriculture, 1947–8.
5. Ashby, E. (1948). "Genetics in the Soviet Union." *Nature*, 162, 912.
6. Bacon, Francis. (1605). *The Advancement of Learning*.
7. Bacon, Francis. (1620). *Novum Organum*.
8. Baker, J. R. (1942). *The Scientific Life*. George Allen & Unwin Ltd., London.
9. Baker, J. R. (1945). *Science and the Planned State*. George Allen & Unwin Ltd., London. Permission to quote kindly granted by Dr. J. R. Baker.
10. Bancroft, W. D. (1928). "The methods of research." *Rice Inst. Pamphlet XV*, p. 167.
11. Bartlett, F. (1947). *Brit. med. J.*, Vol. I, p. 835.
12. Bashford, H. H. (1929). *The Harley Street Calendar*. Constable & Co. Ltd., London.
13. Bate-Smith, E. C. (1948). Personal Communication.
14. Bennetts, H. W. (1946). Presidential Address, Report of Twenty-fifth Meeting of the Australian and New Zealand Assoc. for the Advance of Science, Adelaide.
15. Bernard, Claude. (1865). *An Introduction to the Study of Experimental Medicine* (English translation). Macmillan & Co., New York, 1927. Permission to quote kindly granted by Henry Schuman, Inc., New York.
16. Bradford Hill, A. (1948). *The Principles of Medical Statistics*. The Lancet Ltd., London.
17. Bulloch, W. (1935). *J. Path. Bact.*, 40, 621.
18. Bulloch, W. (1938). *History of Bacteriology*. Oxford University Press, London.
19. Burnet, F. M. (1944). *Bull. Aust. Assoc. Sci. Workers*, No. 55.
20. Butterfield, H. (1949). *The Origins of Modern Science, 1300–1800*. G. Bell & Sons Ltd., London.
21. Cannon, W. B. (1913). Chapter entitled "Experiences of a medical teacher" in *Medical Research and Education*. Science Press, New York.

22. Cannon, W. B. (1945). *The Way of an Investigator*. W. W. Norton & Co. Inc., New York. Permission to quote kindly granted by W. W. Norton & Co. Inc., New York, Publishers, who hold the copyright.

23. Chamberlain, T. C. (1890). " The method of multiple working hypotheses." *Science*, 15, 93.

24. Committee, 1944. *Lancet*, Sept. 16th, p. 373.

25. Conant, J. B. (1947). *On Understanding Science. An Historical Approach*. Oxford Univ. Press, London.

26. Cramer, F. (1896). *The Method of Darwin. A Study in Scientific Method*. McClurg & Co., Chicago.

27. Dale, H. H. (1948). " Accident and Opportunism in Medical Research." *Brit. med. J.*, Sept. 4th, p. 451.

28. Darwin, F. (1888). *Life and Letters of C. Darwin*. John Murray, London.

29. Dewey, J. (1933). *How We Think*. D. C. Heath & Co., Boston. Permission to quote kindly granted by D. C. Heath & Co., Boston.

30. Drewitt, F. D. (1931). *Life of Edward Jenner*. Longmans, Green & Co., London. Permission to quote kindly granted by Longmans, Green & Co., London.

31. Duclaux, E. (1896). *Pasteur: Histoire d'un Esprit*. Sceaux, Paris.

32. Dunn, J. Shaw; Sheehan, H. L.; and McLetchie, N. G. B. (1943). *Lancet*, 1, p. 484.

33. Edwards, J. T. (1948). *Vet. Rec.*, 60, 44.

34. Einstein, Albert. (1933). *The Origin of the General Theory of Relativity*. Jackson, Wylie & Co., Glasgow. Permission to quote kindly granted by Jackson, Son & Co., Glasgow.

35. Einstein, Albert. (1933). Preface in *Where is Science Going?* by Max Planck. Trans. by James Murphy. George Allen & Unwin Ltd., London. Permission to quote kindly granted by George Allen & Unwin Ltd., London.

36. Faraday, Michael. (1844). *Philosophical Mag.*, 24, 136.

37. Felix, A. Personal Communication.

38. Fisher, R. A. (1936). " Has Mendel's work been rediscovered? " *Ann. Sci.*, 1, 115.

39. Fisher, R. A. (1935). *The Design of Experiments*. Oliver & Boyd, London.

40. Fisher, R. A. (1938). *Statistical Methods for Research Workers*. Oliver & Boyd, London and Edinburgh.

41. Fleming, A. (1929). *Brit. J. exp. Path.*, 10, 226.

42. Fleming, A. (1945). *Nature*, 155, 796.

43. Florey, H. (1946). *Brit. Med. Bull.*, 4, 248.

44. Foster, M. (1899). *Claude Bernard*. T. Fisher Unwin Ltd., London. Permission to quote kindly granted by T. Fisher Unwin Ltd., London.

45. Frank, P. (1948). *Einstein. His Life and Times.* Jonathan Cape Ltd., London.
46. Gatke, H. (1895). *Heligoland as an Ornithological Observatory.* D. Douglas, Edinburgh.
47. George, W. H. (1936). *The Scientist in Action. A Scientific Study of his Methods.* Williams & Norgate Ltd., London. Permission to quote kindly granted by Williams & Norgate Ltd., London.
48. Gregg, Alan. (1941). *The Furtherance of Medical Research.* Oxford University Press, London, and Yale University Press. Permission to quote kindly granted by Oxford University Press.
49. Greiff, D., Pinkerton, H., and Moragues, V. (1944). *J. exp. Med.,* **80,** 561.
50. Hadamard, Jacques. (1945). *The Psychology of Invention in the Mathematical Field.* Oxford University Press, London.
51. Harding, Rosamund E. M. (1942). *An Anatomy of Inspiration.* W. Heffer & Sons Ltd., Cambridge. Permission to quote kindly granted by W. Heffer & Sons Ltd., Cambridge.
52. Herter, C. A. Chapter entitled " Imagination and Idealism " in *Medical Research and Education.* Science Press, New York.
53. Hirst, G. K. (1941). *Science,* **94,** 22.
54. Hughes, D. L. (1948). " The present-day organisation of veterinary research in Great Britain: Its Strength and Weaknesses." *Vet. Rec.,* **60,** 461.
55. Kapp, R. O. (1948). *The Presentation of Technical Information.* Constable & Co., London.
56. Kekulé, F. A., quoted by J. R. Baker (1942) from Schutz, G. 1890. *Ber. deut. chem. Ges.,* **23,** 1265.
57. Koch, R. (1890). " On Bacteriology and its Results." Lecture delivered at First General Meeting of Tenth International Medical Congress, Berlin. Trans. by T. W. Hime. Baillière, Tindall & Cox, London.
58. Koenigsberger, L. (1906). *Hermann von Helmholtz.* Trans. by F. A. Welby. Clarendon Press, Oxford. Permission to quote kindly granted by Clarendon Press, Oxford.
59. Lehman, H. C. (1943). " Man's most creative years: then and now." *Science,* **98,** 393.
60. McClelland, L., and Hare, R. (1941). *Canad. Publ. Health J.,* **32,** 530.
61. Mees, C. E. Kenneth, and Baker, J. R. (1946). *The Path of Science.* John Wylie & Sons, New York, and Chapman & Hall Ltd., London.

62. Metchnikoff, Elie, quoted by Fried, B. M. (1938). *Arch. Path.,* **26**, 700. Permission to quote kindly granted by the American Medical Association.
63. Nicolle, Charles. (1932). *Biologie de l'Invention.* Alcan, Paris.
64. North, E. A. Personal Communication.
65. Nutman, P. S., Thornton, H. G., and Quastel, J. H. (1945). *Nature,* **155**, 498.
66. Nuttall, G. H. F. (1938). In *Background to Modern Science,* edited by Needham & Pagel. Cambridge University Press. Permission to quote kindly granted by Cambridge University Press.
67. Ostwald, W. (1910). *Die Forderung der Tages.* Leipzig.
68. Pavlov, I. P. (1936). "Bequest to academic youth." *Science,* **83**, 369. Permission to quote kindly granted by the American Assoc. for the Advancement of Science, Washington.
69. Pearce, R. M. (1913). In *Medical Research and Education.* The Science Press, New York.
70. Planck, Max. (1933). *Where is Science Going?* Trans. by James Murphy. George Allen & Unwin Ltd., London. Permission to quote kindly granted by George Allen & Unwin Ltd., London.
71. Platt, W., and Baker, R. A. (1931). "The Relationship of the Scientific 'Hunch' Research." *J. chem. Educ.,* **8**, 1969.
72. Poincaré, H. (1914). *Science and Method.* Thos. Nelson & Sons, London. Trans. by F. Maitland. Permission to quote kindly granted by Thos. Nelson & Sons, London.
73. Robertson, O. H., Bigg, E., Puck, T. T., and Miller, B. F. (1942). *J. exp. Med.,* **75**, 593.
74. Robertson, T. Brailsford. (1931). *The Spirit of Research.* Preece and Sons, Adelaide.
75. Robinson, V. (1929). *Pathfinders in Medicine.* Medical Life Press, New York.
76. *Rockefeller Foundation Review* for 1943 by R. B. Fosdick.
77. Rous, P. (1948). "Simon Flexner and Medical Discovery." *Science,* **107**, 611.
78. Roux, E., quoted by Duclaux, E. 1896.
79. Russell, Bertrand. (1948). *Human Knowledge. Its Scope and Limits.* George Allen & Unwin Ltd., London.
80. Schiller, F. C. S. (1917). "Scientific Discovery and Logical Proof." In *Studies in the History and Method of Science,* edited by Charles Singer. Clarendon Press, Oxford. Permission to quote kindly granted by Clarendon Press, Oxford.
81. Schmidt, J. (1898). *Vet. Rec.,* **10**, 372.
82. Schmidt, J. (1902). Ibid., **15**, 210, 249, 287, 329.
83. Scott, W. M. (1947) *Vet. Rec.,* **59**, 680.

BIBLIOGRAPHY

84. Sinclair, W. J. (1909). *Semmelweis, His Life and Doctrine.* Manchester University Press.
85. Smith, Theobald. (1929). *Am. J. Med. Sci.,* **178**, 740.
86. Smith, Theobald. (1934). *J. Bact.,* **27**, 19.
87. Snedecor, G. W. (1938). *Statistical Methods applied to Experiments in Agriculture and Biology.* Collegiate Press Inc., Ames, Iowa.
88. Stephenson, Marjory. (1948). " F. Gowland Hopkins." *Biochem. J.,* **42**, 161.
89. Stephenson, Marjory. (1949). *Bacterial Metabolism.* Longmans, Green & Co., London.
90. Taylor, E. L. (1948). "The Present-day Organisation of Veterinary Research in Great Britain: Its Strength and Weaknesses." *Vet. Rec.,* **60**, 451.
91. Topley, W. W. C., and Wilson, G. S. (1929). *The Principles of Bacteriology and Immunity.* Edward Arnold & Co., London.
92. Topley, W. W. C. (1940). *Authority, Observation and Experiment in Medicine.* Linacre Lecture. Cambridge University Press. Permission to quote kindly granted by the Syndics of the Cambridge University Press.
93. Trelease, S. F. (1947). *The Scientific Paper; How to Prepare it; How to Write it.* Williams & Wilkins Co., Baltimore.
94. Trotter, W. (1941). *Collected Papers of Wilfred Trotter.* Oxford University Press, London. Permission to quote kindly granted by Oxford University Press, London.
95. Tyndall, J. (1868). *Faraday as a Discoverer.* Longmans, Green & Co., London.
96. Ungar, J. (1943). *Nature,* **152**, 245.
97. Vallery-Radot, R. (1948). *Life of Pasteur.* Constable & Co. Ltd., London.
98. Wallace, A. R. (1908). *My Life.* Chapman & Hall Ltd., London.
99. Wallas, Graham. (1926). *The Art of Thought.* Jonathan Cape Ltd., London.
100. Walshe, F. M. R. (1944). "Some general considerations on higher or post-graduate medical studies." *Brit. med. J.,* Sept. 2nd, p. 297.
101. Walshe, F. M. R. (1945). "The Integration of Medicine." *Brit. med. J.,* May 26th, p. 723.
102. Warington, K. (1923). *Ann. Bot.,* **37**, 629.
103. Wertheimer, M. (1943). *Productive Thinking.* Harper Bros., New York.
104. Whitby, L. E. H. (1946). *The Science and Art of Medicine.* Cambridge University Press.
105. Willis, R. (1847). *The Works of William Harvey, M.D.* The Sydenham Society, London.

106. Wilson, G. S. (1947). *Brit. med. J.*, Nov. 29th, p. 855.

107. Winslow, C. E. A. (1943). *The Conquest of Epidemic Diseases.* Princeton University Press.

108. Zinsser, Hans. (1940). *As I Remember Him.* Macmillan & Co. Ltd., London; Little, Brown & Co., Boston; and the Atlantic Monthly Press. Permission to quote kindly granted by the publishers.

109. Gram, C. (1884). *Fortschritte der Medicin, Jahrg.* II, p. 185.

110. Cajal, S. Ramon y (1951). *Precepts and Counsels on Scientific Investigation, Stimulants of the Spirit.* Trans by J. M. Sanchez-Perez. Pacific Press Publ. Assn., Mountain View, California.

111. Conant, J. B. (1951). *Science and Commonsense.* Oxford University Press.

112. Dubos, René J. (1950). *Louis Pasteur: Freelance of Science.* Little, Brown & Co., Boston.

113. Marquardt, M. (1949). *Paul Ehrlich.* Wm. Heinemann Ltd.

114. Peters, J. T. (1940). *Act. med. Scand.*, **126**, 60.

115. Parkes, A. S. (1956). Proceedings of the III International Congress on Animal Reproduction, Cambridge, 25–30 June, 1956.

INDEX

ACKNOWLEDGEMENTS

For their kind permission to reproduce paintings and photographs in this book, the Author wishes to thank the following:

The National Portrait Gallery, for MICHAEL FARADAY and EDWARD JENNER.

The Royal Society, for SIR F. GOWLAND HOPKINS.

The Director of the Pasteur Institute, Paris, for LOUIS PASTEUR.

Messrs. Macmillan and Co., Ltd., for THOMAS HUXLEY (from *Memoirs of Thomas Huxley*, by M. Foster).

Messrs. Allen and Unwin, Ltd., for GREGOR MENDEL (from *Life of Mendel*, by Hugo Iltis).

Picture Post, for CLAUDE BERNARD.

Harper's Magazine, for CHARLES DARWIN.

Martha Marquardt, for PAUL EHRLICH (from her *Paul Ehrlich*, published by Heinemann).

The editor, *The Journal of Pathology*, for THEOBALD SMITH.

Mrs. W. B. Cannon, for WALTER B. CANNON.

Messrs. J. Russell and Sons, for SIR HENRY DALE and SIR HOWARD FLOREY.

Topical Press, for SIR ALEXANDER FLEMING.

Lotte Meitner-Graf, for MAX PLANCK.

CPSIA information can be obtained at www.ICGtesting.com
Printed in the USA
BVOW030623300413

319425BV00001B/1/A

9 781932 846058